I0070093

Our Cosmic Ancestry: Origins, Evolution, Metamorphosis of Life From Other Planets

Alien Viruses, Extraterrestrial Bacteria, Interplanetary Gene Transfer, Genetics, Metamorphosis, Cambrian Explosion: Humans

Rhawn Gabriel Joseph, Ph.D.
Cosmology.com

Evolution on Earth is the Replication and Metamorphosis of Extraterrestrial Life From Other Worlds

"The Earth was genetically seeded to grow complex life. When the descendants of some of these microbes fell to Earth, they possessed the genetic libraries and the genetic information for replicating life forms which long ago evolved on other worlds. What has taken place on this planet, during the course of the last 4.6 billion years leading to the Cambrian Explosion and then culminating in the evolution of woman and man, is not a random evolution, but the replication, metamorphosis, and evolution of alien life forms from other planets."

Cosmology Science Publishers

ISBN/10: 1-938024-54-0
ISBN/13: 978-1-938024-54-2

Contents

7. Evolutionary Metamorphosis, Embryogenesis, and the DNA-Supra Organism 116

8. Multi-Regional Human Metamorphosis   127

8. Evolution in the Ancient Corners of the Cosmos: Even the Gods Have Gods Who Have Gods 163

1. Overview: The Evolution of Alien Life From Other Planets

"If life were to suddenly appear on a desert island we wouldn't claim it was randomly assembled in an organic soup or created by the hand of god; we'd conclude it washed to shore or fell from the sky. The Earth too, is an island, orbiting in a sea of space, and living creatures and their DNA have been washing to shore and falling from the sky since our planets creation" --Joseph (2000).

The discoveries and theories detailed in this text can be summarized as follows: (**1**) Life on Earth has extraterrestrial origins. (**2**) These extraterrestrials are descendants of prokaryotes and eurkayotes which had long ago "fallen" unto other planets where they engaged in interplanetary horizontal gene exchange. (**3**) The genetic information for the evolution of life on Earth has extraterrestrial origins, (**4**) Once on Earth extraterrestrial genes, including "silent genes" and the genetic material for the generation of additional genes, were transferred into the eukaryotic genome from viruses and bacteria which "fell" to Earth and (**5**) and their DNA contained all the genetic information necessary for the evolution of every species which has walked, crawled, swam or slithered across this planet--including species which never evolved on Earth. (**6**) Over the course of evolution on Earth, various species with their "libraries" of extraterrestrial genetic information, genetically and biologically altered the environment via the liberation or secretion of oxygen and calcium and other substances which (**7**) acted on gene selection leading to new species adapted to a world which had been genetically engineered for their survival. (**8**) When sufficient oxygen, calcium etc. had been secreted, and with the formation of the ozone layer, silent and other genes were activated giving rise to the Cambrian explosion (**9**) and the evolution of the eye, brain and skeletal system. (**10**) Bacteria, archae, and viruses, and perhaps even more complex species, continually fall from space and (**11**) continue to insert genes, via horizontal gene transfer, into the eukaryotic genome. (**12**) All these factors explain the step-wise, punctuated equilibrium which characterizes evolution on this planet, leading to the multi-regional evolution and metamorphosis of numerous species of human which indicates (**13**) Earth was genetically seeded to "grow" every species which has appeared on Earth culminating in woman and man (**14**) according to the genetic principles of "evolutionary metamorphosis."

Alien Viruses, Genetics, Cambrian Explosion: Humans

Summarizing Quotes

"Life on Earth did not emerge from an organic soup or undersea thermal vent. It fell from the sky encased in all manner of stellar debris which pounded the planet for the first 700 millions years after the creation." --Joseph (1997, 2000)

"The genetic seeds of life swarm throughout the cosmos, and some of these genetic "seeds" fell to Earth, as well as on other planets. And these genetic "seeds" contained the instructions for the metamorphosis of all life, including woman and man." --Joseph (2000).

"And these "genetic seeds" of life contained the DNA/genetic instructions for the tree of life, and for fashioning and evolving every multi-cellular creature which has walked, crawled, swam, or slithered upon the Earth --Joseph (2000, 2009)

"DNA acts to purposefully modify the environment, which acts on gene selection, so as to fulfill specific genetic goals: the dispersal and activation of silent DNA and the replication of life forms that long ago lived on other planets." --Joseph (2000).

"Just as DNA contains the genetic instructions for creating an embryo, fetus, baby, child, adolescent and adult, and the genetic instructions for inducing the metamorphosis and transformation of a caterpillar into a butterfly, the "seeds of life swarming throughout the cosmos" and their DNA, have also contained the instructions for genetically engineering the environment. The environment has been genetically engineered so as to make possible the step-wise creation, expression, and dispersal of all manner of DNA-based life forms, including, on this planet and innumerable other ancient worlds, the likes of woman and man." --Joseph (2000).

"Just as the transformation of an embryo into an infant, or a caterpillar into a butterfly is genetically coded and not a function of mutation, random factors, or "coincidence," the stepwise progression from simple to complex species is genetically coded." --Joseph (2000).

"Again, however, these "new" species and "new" products, are preprogrammed into the genetic code, and represent not the evolution of new species, but the metamorphosis and replication of species who long ago dwelled on other planets." --Joseph (2000).

"Rather than taking just 9 months, or a single season, these changes and the metamorphosis of "new" species takes billions of years. And, as is evident on Earth, and as can be deduced from the DNA organization and expression and the progressive emergence of increasingly complex and intelligent species, as the environment was genetically engineered, this has led not just to diversity, but the replication of previous life forms that long ago lived on other planets, including fish, frogs, reptiles, repto-mammals, cats, dogs, and woman and man." --Joseph (2000).

"What has been called a random evolution has been under precise genetic regulatory control. Genes are not randomly expressed, nor do they randomly evolve. They are inherited and their expression is highly regulated. Further, these genes were acquired through horizontal gene transfer from extra-terrestrial sources when space-journeying viruses, bacteria, and archae were cast from planet to planet, from solar system to solar system, and from galaxy to galaxy (Joseph and Schild 2010b). The first Earthlings, and their viral genetic luggage, already possessed all the necessary genes for genetically engineering the biosphere and for generating every life form which evolved on Earth. What has been called evolution can be likened to embryological development and is a form of metamorphosis: the replication of complex creatures that long ago lived on other planets." -Joseph, R. (2009)

"Thus, what has been described as a random evolution, is in fact under the interactive control of genetic and biologically altered environmental influences which directly impact genetic mechanisms involved in gene silencing, gene duplication, and gene expression, thereby giving rise to traits, functions, organs, and species, which had been precoded into silent genes inherited from ancestral species. These genes and regulatory elements were donated to the eukaryotic genome by viruses and prokaryotes--the ancestors of which, arrived on Earth from other, more ancient worlds.Evolution is not random. Evolution is embrogenesis and metamorphosis: The replication of life forms which long ago lived on other planets."--Joseph, R. (2009)

"Prokaryotes can journey from world to world, exchange and acquire genes, and act with yet other microbes to induce environmental changes to activate these genes, thus guiding the metamorphosis and replication of creatures that long ago lived on other planets.The genetic seeds of life flow throughout the cosmos, and identical genetic seeds have fallen upon innumerable worlds, including those much older than our own. The Earth was genetically seeded to grow complex life, and what has taken place on this planet, during the course of the last 4.6 billion years, is not a random evolution, but the replication, metamorphosis, and evolution of life from other planets." --Joseph, R. (2009)

"The first microscopic life forms to take up residence on this world included archae, bacteria, and viruses, and these microbes possessed the genes, genetic mechanisms, "silent genes," and the necessary genetic material and instructions for the evolution and metamorphosis of all life including those which have never, or have not yet evolved on Earth.-- Joseph (2013)

"Any planet with oceans, atmosphere, and surface dwelling organisms will inevitably seed surrounding moons and planets with microbes and possibly eukaryotic life. Microbial organisms from a single source may even come to be distributed on a galaxy-wide scale. Because dispersal and contamination is ongoing, eventually the descendants of these original sojourners from the stars would be hurled back and forth between planets and solar systems and come into contact and exchange DNA with their microbial "cousins" via horizontal gene transfer (Joseph 2000, 2009b,c). When the descendants of some of these microbes fell to Earth, they possessed the genetic libraries and the genetic information for replicating life forms which long ago evolved on other worlds --Joseph & Schild, (2010)

"However, although over 40% of the human genome consists of genes inserted by Viruses, including genes coding for the human brain, much of the remainder of the human genome can be traced to genes inserted by Archae and Bacteria (Joseph 2009b,c). Thus, evolution leading to humans, has been guided by genes inserted by microbes and Viruses whose own ancestry can be traced to life forms which journeyed here from other planets. This indicates that similar genetic interactions leading to similar evolutionary progressions must have taken place on other planets.An analysis of the microbe, viral, and eukaryotic genome and the evolutionary progression which has taken place on this biologically engineered planet leads to this conclusion: The first viruses and life forms to arrive on Earth contained the genetic instructions for creating all of life, and some of these genes were transferred to or gave rise to the eukaryotic genome. Just as an apple seed contains the genetic instructions for the development of an apple tree, these genetic seeds of life contained the genetic instructions for the tree of life, and for every creature which has walked, crawled, swam, or slithered across the Earth." --Joseph & Schild, 2010

"Genes act on the environment, and the biologically altered environment acts on gene selection, thereby expressing traits which had been encoded into genes acquired from life on other planets.Evolution on Earth could be likened to metamorphosis and embryology. Metamorphosis is genetically regulated. All aspects of development are guided and controlled by genetic-environmental interactions. Embryogenesis is under genetic control. Why should evolution be any different? However, rather than 9 months, or a single season, it takes billions of years to grow a human from a single cell.What has been called "evolution" is under ge-

Origins, Evolution, Metamrphosis of Life From Other Planets

netic regulatory control, in coordination with the biological activity of innumerable life forms which genetically engineer the environment. Genes act on genes, genes act on the environment, and the altered environment acts on gene selection, thereby giving rise to an evolutionary progression from simple cell to sentient intelligent being, each evolving into a world which has been genetically prepared for them." --Joseph & Schild, 2010

What has taken place on Earth represents not a random evolution, but the metamorphosis and replication of living creatures which long ago lived on other planets. --Joseph 2000

"Our ancient ancestors, and their genes, journeyed here, from the stars."

10

2. Origins and Evolution of Alien Life From Other Planets

"If life were to suddenly appear on a desert island we wouldn't claim it was randomly assembled in an organic soup or created by the hand of god; we'd conclude it washed to shore or fell from the sky. The Earth too, is an island, orbiting in a sea of space, and living creatures and their DNA have been washing to shore and falling from the sky since our planets creation" (Joseph, 2000a).

Origins and Evolution of Life From Space: Metamorphosis

A wealth of data from microbiology, genetics, chemistry, and astrobiology, demonstrates that the first living creatures to take root on Earth must have been deposited on this planet encased in meteors, asteroids, comets, and oceans of ice; that the ancestry of these first Earthlings leads to viruses and extraterrestrial Prokaryotes and Eukaryotes which evolved on other planets, and to genes whose origins extends interminably into the long ago.

Life on Earth came from other planets. And then it began to evolve according to the same genetic principles which govern embryology and metamorphosis.

There is considerable genetic evidence which indicates the first life forms to take up residence on this world were accompanied by viruses. And these microbes and viruses possessed the genes, genetic mechanisms, "silent genes," and the necessary genetic instructions for the evolution and metamorphosis of all life including those which have never, or have not yet evolved on Earth (Joseph 1997, 2000a, 2009a,b, 2010a). After arriving on Earth, these genes were inserted into what became the multicellular Eukaryotic genome.

These first Earthlings included Cyanobacteria which along with other Prokaryotes also labored to terraform the planet, liberating a variety of gasses, elements, and minerals, including and especially oxygen and calcium. The changing environment acted on gene selection, expressing a variety of genes, including "silent genes" which encoded, for example, the genetic instructions for the metamorphosis of eyes, bones, and brains. A variety of Prokaryotes, most notably Cyanobacteria, have continually biologically altered the environment which has acted on the genes these same species had transferred into the Eukaryotic genome, thereby producing evolutionary change. It is also precisely because of the lengths of time necessary to alter the biosphere sufficiently so to build up these gasses, minerals, etc., that once certain critical levels were reached long periods of evolutionary stasis were punctuated by explosive bursts of speciation and dramatic morphological change, often accompanied by extinctions.

What has been traditionally referred to as "evolution" is effected by the changing environment which acts on and expresses various genes, and is under genetic regulatory control, similar in many ways to the genetic mechanisms governing embryology and metamorphosis. Embryological development and metamorphosis are genetically regulated and effected by the internal and external environment, and the same is true of evolution (Joseph 1997, 2000a, 2009d).

The facts and theories on the evolution and origins of life, as detailed here, are supported by a wealth of data from genetics, astrobiology, microbiology, virology and the fossil records. Further, the data on evolution can be interpreted in two ways: **1)** Archae, bacteria, and viruses have repeatedly inserted genes into the Eukaryotic genome thereby influencing the direction and trajectory of life's evolution on this planet; and that viruses, Prokaryotes, and Eukaryotes interact genetically, not as independent entities, but in a coordinated, choreographed fashion reminiscent of the purposeful and genetically regulated interactions of those cells comprising a complex multi-cellular supra-organism. **2)** Extraterrestrial viruses, archae, and bacteria have repeatedly inserted genes into the Eukaryotic genome including regulatory genes which have governed gene expression, and they contained all the genes, genetic elements, and genetic instructions for altering the environment which also acts on those genes donated to Eukaryotes by viruses and Prokaryotes; and what has been called *evolution* is genetically regulated and a form of embryogenesis and metamorphosis: the replication of creatures who long ago lived on other planets.

The Death of Darwinism: Small Steps vs Evolutionary Leaps

The evidence which will be presented in this text is consistent with and supported by genetics, the evolution of cellular structures such as proteins, and the fossil record which demonstrates that the evolution of species is characterized by long periods of evolutionary stasis followed by quantum evolutionary leaps to subsequent species without benefit of intermediate forms (Bose 2013; Gould, 2002; Ingles-Prieto, et al. 2013; Rabosky 2012). That is, there appear to be periods of little or no evolutionary change which are punctuated by rapid, sudden and dramatic change (Bose 2013; Ingles-Prieto, et al. 2013; Rabosky 2012); a phenomenon that Eldredge and Gould (1972; Gould 2002) described as "punctuated equilibrium."

Quantum leaps of frenzied evolutionary and genetic regulatory activity, including increases in genomic complexity followed by stasis has also characterized the genomic interactions of Prokaryotes around the time they ushered in the evolution of multicellularity in Eukaryotes. According to Wolf and Koonin (2013) "Quantitatively, the evolution of genomes appears to be "punctuated by episodes of complexification" and it is this "explosive, innovation phase that leads to an abrupt increase in genome complexity." However, in Prokaryotes this was "followed by a much longer reductive phase," stasis, "which encompasses either a neutral ratchet of genetic material loss or adaptive genome streamlining." Gene loss, however, is not a function of stasis, but is related to evolutionary apoptosis (or "extinction" at the species level) and a function of the transfer of Prokaryote genes to the genomes of Eukaryotes and viruses (Joseph 2009b,e)

It is recognized that most species have become extinct and some species show little or no evidence of evolution whereas others show giant leaps to subsequent forms. Consider, for example, sharks, lungfishes, and coelacanths which are considered "living fossils." Many of these species appeared over 400 million years ago and are characterized by low species diversity and have displayed morphological stasis for much of their evolutionary history. In fact, many of the older more ancient lineages (which have not gone extinct) undergo little morphological change and low rates of species diversification relative to more recent lineages (Gingerich 2001, Rabosky et al. 2012).

Yet other species display relatively brief periods of morphological stasis which is followed by rapid periods of diversification into a variety of functional forms and a vast number of species. Speciation and morphological evolution are linked and consistent with a punctuated equilibrium model of evolution; and this association can be attributed to genetics; that is, at the genetic level, some species have a greater phenotypic evolvability, a greater genetic capacity to evolve (Adamowicz et al., 2008; Pigliucci, 2008; Rabosky 2012; Rabosky, et al. 2013). Evolvability, therefore, is a function of genetic potential and has little to do with Darwinism or Darwin's concepts of small steps.

Evolvability, and these quantum evolutionary leaps are not due to chance but are related to biological alterations of the environment, and complex interactions involving the rearrangement and duplication of genes and the entire genome, and viral gene and regulatory gene activity which activates genes donated by Prokaryotes to the Eukaryote genome including silent genes which code for advanced characteristics such as hearts, eyes, bones and brains (Joseph 2000a, 2009d,e). It is these silent genes and these genetic, viral, environmental interactions, which explain, for example, why the genes coding for these structures were inherited from ancestral forms and appeared in the genomes of heartless, boneless, brainless "living fossils" around 640 mya (Sakarya et al., 2007; Srivastava et al., 2008), and why after the passage of another hundred millions years, and following the biological production and buildup of oxygen and calcium to critical levels, so many species suddenly evolved hearts, eyes, bones and brains, with the onset of the Cambrian Explosion 540 mya, and with no evidence of intermediate or transitional forms.

Contrary to Darwin's theory, there is no fossil evidence of gradual change from one species to another or any fossil record of transitional species acting as an evolutionary bridge between species (Eldredge & Gould 1972; Gould 2002). Evolution, as demonstrated by the fossil record and genetics, often occurs in leaps involving major changes in structure and function, and thus the fossil record refutes Darwin's theory which emphasizes "infinitesimally small inherited modifications" (Darwin, 1866).

Darwin Attributes Evolution to the Powers of The Creator

Darwin has been elevated to near deity status and considerable effort has been extended to obfuscate, obscure, dismiss, and explain away his belief in a "God" or that he trained to be a minister of religion. Nevertheless, Darwin, in his book "Origins of Species" repeatedly refers to "god" to a "creator" to a "spirit" and to unexplained quasi-supernatural "powers" as being responsible for evolution and "natural selection." Darwin refers to "The Creator" and "God" eleven times in his book. Darwin (1860, 1872) explains that "the Creator provided the breath of life" and that "There is grandeur in this view of life, with its several powers, having been originally breathed by the Creator into a few forms or into one."

God, the Creator, or these supernatural "powers" according to Darwin (1859), are also responsible for guiding evolution and natural selection by actively choosing between the good and the bad: "To my mind it accords better with what we know of the laws impressed on matter by the Creator" (Darwin 1859)

Darwin repeatedly defends his God The Creator. And to those who reject the Creator's involvement in evolution by claiming that species suddenly appeared or evolved without god's help, Darwin (1860) says such beliefs are anathema: "To admit this view is, as it seems to me, to reject a real for an unreal, or at least for an unknown, cause. It makes the works of God a mere mockery and deception." Darwin also complained that the minds of mortal men are not capable of appreciating God's power: "Have we any right to assume that the Creator works by intellectual powers like those of man?"

Darwin also gives "natural selection" all seeing infallible powers; a benevolent omniscience which makes god-like judgements about good and evil: "I think it can be shown that there is such an unerring power at work in Natural Selection (the title of my book), which selects exclusively for the good of each organic being" (Darwin 1853, 1857, 1858). "Natural selection is daily and hourly scrutinising, throughout the world, the slightest variations; rejecting those that are bad, preserving and adding up all that are good; silently and insensibly working, whenever and wherever opportunity offers, at the improvement of each organic being in relation to its organic and inorganic conditions of life" (Darwin 1866).

As summed up by Darwin (1838): "There is one living spirit prevalent over this world, (subject to certain contingencies of organic matter & chiefly heat), which assumes a multitude of forms each having acting principle according to subordinate laws...." And what is that "one living spirit"? According to Christian theology, that spirit is God.

Darwin (1892) was not a scientist, but trained to be a minister of religion, to preach the word of Jesus Christ and his god, and he began his career hoping to make the Bible scientific. Darwin even admitted a willingness to "invent evidence" to support his belief that the Biblical stories in the Gospels were factual and that

Jesus Christ was God: "I was very unwilling to give up my belief;—I feel sure of this for I can well remember often and often inventing day-dreams of old letters between distinguished Romans and manuscripts being discovered at Pompeii or elsewhere which confirmed in the most striking manner all that was written in the Gospels" (Darwin, 1892).

Not surprisingly, although later in his life he claimed to be an atheist, Darwin's theories about evolution were heavily influenced by his religious beliefs; so much so, that even Darwin (1857, 1859) admitted that he feared his speculations had little scientific basis and bordered on fantasy: "I am quite conscious that my speculations run quite beyond the bounds of true science." "Often a cold shudder has run through me, and I have asked myself whether I may not have devoted my life to a phantasy."

However, even if we ignore Darwin's repeated appeals to supernatural forces and the powers of his God the Creator, the fact remains that Darwin's theory is refuted by genetics, the fossil record, and the considerable body of evidence which indicates that evolution is characterized by great evolutionary leaps punctuated by periods of stasis. Darwin (1859) himself warned: 'Natural selection can act only by the preservation and accumulation of infinitesimally small inherited modifications, each profitable to the preserved being ... If it could be demonstrated that any complex organ existed, which could not possibly have been formed by numerous, successive, slight modifications, my theory would absolutely break down."

The Cambrian Explosion, 540 million years ago, when hundreds of species belonging to dozens of phyla suddenly acquired bones, brains, and modern eyes without intermediate forms, completely refutes Darwin's strict gradualism which is an essential staple of his theory.

Darwin's speculations, as he admits, are not wholly scientific, and are based on historical reconstructions. However, a scientific theory must make predictions and be testable. It is impossible to make scientific predictions or to scientifically test any aspect of Darwin's theory; a theory which is based on hindsight and allusions to metaphysical and supernatural powers wielded by a god, a creator.

Darwin's entire theory is in fact a tautology based on hindsight, judgments after the fact, and circular reasoning which can be summed up as: "The fit survive because they are fit. Survivors are fit because they survive." Given that over 99.9% of all species have become extinct (Prothero (1998), then, by Darwin's definition, because almost no species survives then almost no species are fit.

The Genetic Seeds of Life

There is a wealth of evidence which supports a theory that "evolution" has taken place on this planet. However, there is little or no data which substantiates and considerable evidence which utterly refutes Darwin's speculations about evolution.

By contrast, and as will be detailed, there is considerable data from genetics, microbiology, virology, and the fossil record which demonstrates that life could not have begun on Earth; that the first and subsequent species to arrive on this planet contained within their genomes all the necessary genetic material for the evolution of all subsequent species; and that Prokaryotes, Cyanobacteria in particular, have altered the biosphere which in turn acted on genes transferred to Eukaryotes by viruses and Prokaryotes thereby giving rise to a succession of species and quantum leaps in evolution followed by periods of stasis and species eradication; a pruning of the tree of life. And, as will be discussed, there is experimental evidence demonstrating that microbes and viruses can survive a journey through space and a crash landing onto the surface of a planet, that these species constantly transfer genes to each other and to the Eukaryotic genome, and that "silent genes" can be experimentally activated or expressed by varying the environment.

Extraterrestrial Prokaryotes and viruses, these "genetic seeds of life" which swarm throughout the cosmos, contain all the genetic information for the evolution of all life, of which life on Earth is merely a sample. Evolution is not random, but is a form of metamorphosis and embryology, the replication of species who long ago evolved on other worlds.

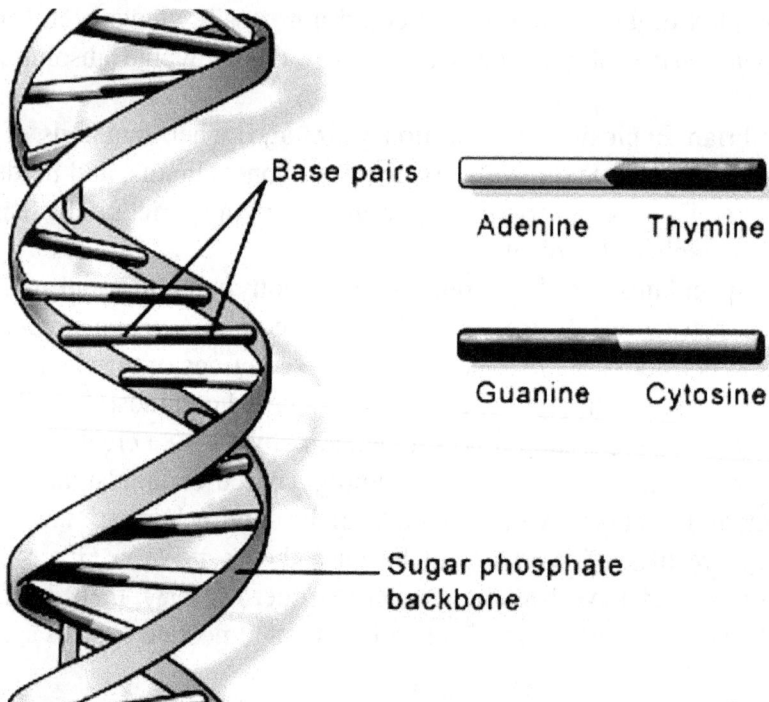

Base pairs

Adenine   Thymine

Guanine   Cytosine

Sugar phosphate backbone

Life Could Not Have Begun On Earth

Life on Earth was present from the very beginning. Coupled with other data, this can only mean that these first Earthly life forms, these "genetic seeds of life" came from other planets and were deposited here contained within meteors, comets, asteroids and extraterrestrial debris (Joseph 2000a; Joseph & Schild 2010a,b).

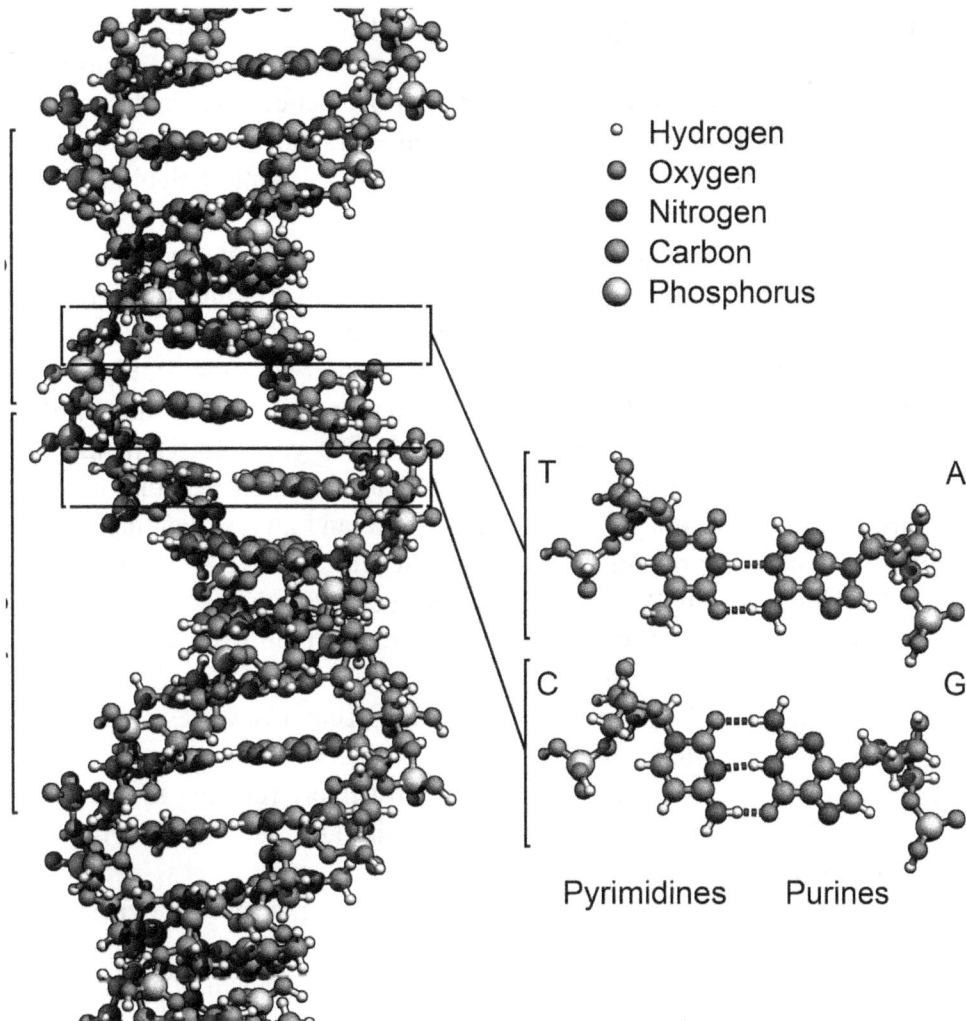

By contrast, there is no evidence that life originated on Earth and considerable data which refutes this possibility. The problems with an Earth-centered abiogenesis have been detailed and reviewed by Joseph and Schild (2010a) and can be summed up as follows: **A)** Life was present on Earth almost from the beginning. **B)** There is no evidence that life has been or can be produced from non-life on this planet. **C)** Free oxygen, phosphorus, sugar, boron, molybdenum, and other essential ingredients for creating proteins, nucleotides, DNA, RNA, and life, were not freely available on the new Earth. **D)** Naked DNA and complex organic molecules would have been destroyed by the constant thermal motion, radiation, and the hellish environment of the early Earth. **E)** Genetic analyses indicates it would take from 10 billion to 14 billion years for a single gene or a genome consisting of just one gene, to undergo repeated duplicative events so as to fashion a minimal gene set necessary for sustaining and maintaining life (Joseph & Wickramasinghe 2011; Sharon & Goron 2013). However, this genetic analysis did not include the length of time necessary to fashion the first protein, the first nucleotides, the first DNA macro-molecule, or the first cellular membranes. **F)** Statistically, the creation of the simplest of single celled organisms and all its components would have required over a trillion years and over a trillion chance combinations of all the right ingredients (Hoyle, 1974; Joseph & Schild, 2010a).

Hellish Hadean Earth: Biological Destruction, Few Ingredients for Life

For almost 800 million years Earth was continually pounded by meteors, asteroids, comets, and mountain-size, moon-size, and even planet-sized debris (Belbruno & Gott, 2005; Jacobsen, 2005; Poitrasson et al. 2004, Rankenburg et al. 2006; Schoenberg et al. 2002). As there is evidence of biological activity in this planet's oldest rocks dated to 4.28 and 4.2 billion years go (Nemchin et al. 2008; O'Neil et al. 2008), and given the well substantiated fact that microbes and viruses can withstand a journey through space and a crash landing onto the surface of a planet (Burchell et al. 2004; Burchella et al. 2001; Fekete et al., 2004, 2005; Horneck et al. 2001a.b, Mastrapaa et al. 2001; Nicholson et al. 2004) coupled with discoveries of extraterrestrial microfossils in a variety of meteors some older than this solar system (Claus & Nagy 1961; Hoover, 2011; McKay et al. 1996; Pflug 1984; Nagy et al. 1961,1963a,b; Zhmur & Gerasimenko 1999; Zhmur et al. 1997), including those which resemble virsues, it can be assumed that innumerable microbes and viruses were most likely residing in that debris, which is how Earth became contaminated with life (Joseph 1997, 2000a, 2008, 2009a; Joseph & Schild 2010a,b).

By contrast, the violent, volatile, shattering, shocking, turbulent, hyperthermal conditions on the early Earth, coupled with the lack of a significant atmosphere, extreme temperatures, and continual bathing in gamma, cosmic, and UV rays, would have destroyed all complex organic carbon based molecules including nucleotides and any DNA not already protected by a cell wall; conditions which would have made the assembly of even the most rudimentary life-associated elements an impossibility (Crick 1981; Ehrenfreund & Sephton 2006).

Moreover, the new Earth was radiating radioactive energy and heat producing isotopes such as potassium-40, uranium-238, uranium-235, and thorium-232 (Jordan 1979; Robertson 2001; Turcotte & Schubert 2002). In the early history of Earth, these heat producing isotopes would have been at full strength (Turcotte and Schubert 2002), thereby destroying all biological molecules, proteins, and naked DNA; but not any of the many microbial species and viruses which are radiation resistant.

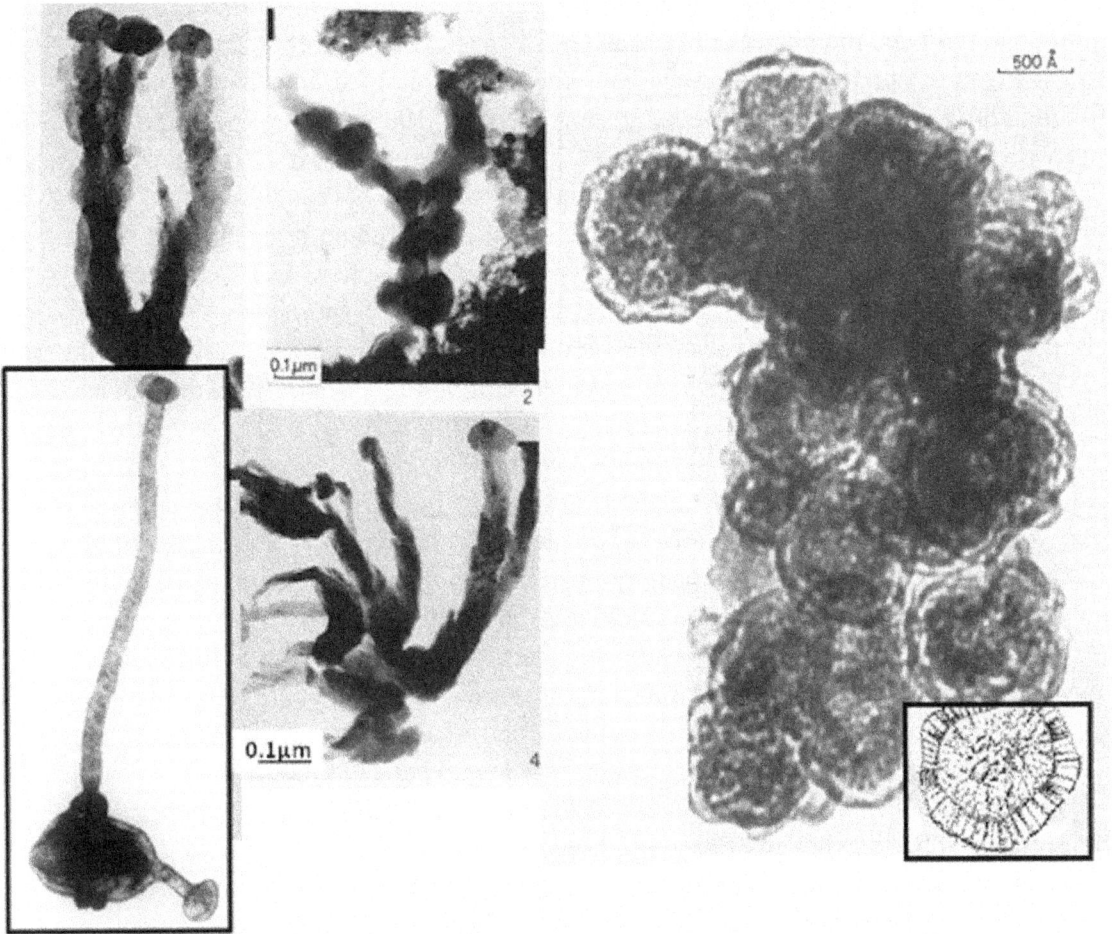

(Left) The comparison of a biological structure in the Murchison meteorite with structures resembling Cyanobacteria and an iron-oxidising microorganism – pedomicrobium. (Right) An electron micrograph of a structure resembling a clump of viruses – influenza virus – also found in the Murchison meteorite. The drawing in the inset is a representation of a modern influenza virus displaying similarities in structure to what appears to be fossil viruses (From Joseph & Wickramasinghe 2010).

For the first billion years after Earth became a member of this solar system, the sun may have been 80% of its current size, 70% as luminous, and did not generate as much heat as the modern sun (Gough 1981; Kasting & Ono 2006; Lang 2001). This relatively feeble heat source, coupled with sunlight blocking debris in the atmosphere secondary to bolide impacts, might be expected to contribute to global cooling. And yet, Earth was not cold, but blistering hot (Kasting & Ackerman 1986). The first period of global warming was triggered by internally generated geothermal heat flow (Davies 1990), the excretion and liberation of heat trapping greenhouse gasses (Joseph 2009d,e) and the tremendous heat generated as stellar debris pounded the planet until around 3.8 bya (Schoenberg et al. 2002).

In addition, heat-trapping gases were being pumped into the atmosphere including carbon dioxide (Kasting & Ackerman 1988; Sleep & Zahnle 2001; Walker 1985) and H_2 secondary to volcanism (Berner 2004; Kirschvink 1992; Hoffman et al. 1998) and microbial activity (Joseph 2009d,e).

The residue following asteroid impact contributed green house gasses. When rock is vaporized following impact, a rock vapor atmosphere and heavy volatiles are produced consisting of carbon dioxide and hydrogen, thereby creating a heat-trapping haze. All surface rocks were continually being vaporized then solidifying and then vaporizing; and yet, there is evidence of biological activity in this planet's oldest rocks, those which were the first to re-solidify and stay intact, dated to 4.28 and 4 bya (Nemchin et al. 2008; O'Neil et al. 2008).

According to Kasting and Ackerman (1986), if the early atmosphere contained 10 bars of CO_2 this would have created a dense heat-trapping greenhouse atmosphere. Coupled with impact induced heat, parts of the planet may have been at the melting point. Others have calculated that atmospheric CO_2 levels were much lower (Sleep & Zahnle 2001) which suggests that the planet was broiling but not melting. Of course, the temperatures of some areas of Earth would have fluctuated due to transient variables such as frequency and size of asteroid and meteor strikes and the presence or absence of oceans of water. Therefore, at varying times for the first 800 million years, pockets of the Hadean Earth may have melted from impact whereas other regions would have been relatively hot but not broiling, whereas at other times the warmer regions became broiling and those areas that had melted becoming cooler and solidifying. The repeated vaporization of surface rocks explains why the oldest rocks are 4.28 and not 4.6 billion years in age.

Nevertheless, despite the evidence of life in these ancient rocks, these conditions would have destroyed all pre-biological molecules, and naked proteins, nucleotides, DNA and RNA. Although hyperthermophiles could surive, life could not have been fashioned under these conditions.

Oceans of water, delivered by comets, were probably also crashing into the new Earth, thus cooling parts of the planet and providing Earth with oceans (Drake 2005) as well as life (Joseph 1996, 2000a). Evidence for massive quantities of water, by 4.2 bya, are indicated indirectly by an analysis of zircons (Valley 2002; Wilde et al., 2001). Zircons crystals are created during rock formation and have been discovered embedded in sedimentary rocks located in the Jack Hills areas of Western Australia. A small fraction have been determined to be 4.2 billion years old and to have crystalized at an average temperature of 690 degrees Celsius, which suggests the presence of water (Valley 2002; Wilde et al., 2001). By contrast rock formation following impact in the absence of water, occurs at around 900 to 1,200 degrees. Thus, because these zicron containing rocks were formed at lower impact temperatures, water must have been present.

Life Was Present From the Beginning: Hyperthermophiles

At the same time rocks were vaporized then hardened, and oceans of water were falling upon the surface of Earth, there is evidence of biological activity (Nemchin et al. 2008; O'Neil et al. 2008), which suggests that microbes were also living in those oceans of water whereas others had been deposited on Earth encased in extraterrestrial debris.

However, at temperatures of 690 degrees Celsius, the oceans of Earth had to be boiling. Presumably, the deepening oceans were prevented from evaporating into space due to Earth's gravity, and the presence of a thick CO_2 atmosphere.

Although biological molecules and proteins would be destroyed at surface temperatures averaging around 80°C (Kasting & Ackerman 1986) and periodically rising to 690°C (Valley 2002; Wilde et al., 2001), numerous species of hyperthermophiles could easily survive and flourish if they were dwelling deep beneath the rising seas or less than a kilometer below Earth's surface. Those survivors could have included the ancestors of *Desulforudis audaxviator*, which lives 2.8-kilometers (1.74 miles) beneath the earth (Joseph 2009a). In fact, microbes have been recovered from geothermally heated rocks and aside 400 °C rock chimneys at the bottom of the ocean (Setter 2002). *Baccilus infernus* can grow and multiply in 100°C boiling water and they can do so without oxygen (Setter 2002) and the continue to thrive even while bathing in temperatures exceeding 117°C (Boone et al. 1995). Moreover, these species can also form spores, protected by heat shock proteins, which enables them to survive at even higher temperatures. Moreover, after forming spores, they can awaken from their long slumber even after 250 million (Vreeland et al. 2000) to 600 million years (Dombrowski 1963).

Since they could not have been fashioned on Earth, then where did these microbes and viruses originate? Microbes and viruses can easily survive a journey through space, including the violent ejection from the surface of a planet, the frigid temperatures and vacuum of an interstellar environment, and the UV rays, cosmic rays, gamma rays, and ionizing radiation they would encounter, and when descending to the surface of a planet the shock waves of a violent impact as the debris they are in strikes the upper atmosphere even if accompanied by high atmospheric explosions (Szewczyk et al., 2005). They can in in fact survive reentry speeds of up to 9700 km h-1 (McLean et al., 2006) and thus the descent and then the crash landing upon the surface of a planet (Burchell et al. 2004; Burchella et al. 2001; Horneck et al. 2001a.b, 1994; Mastrapaa et al. 2001; Nicholson et al. 2000). Since there is no evidence life can be generated from non-life, at least on Earth, it is logical to assume that the ancestry of the first Earthlings leads back into space.

Therefore, for the first few hundred million years new Earth was hellishly hot (hence, the name "Hadean Earth") with surface temperatures averaging around 80°C (176°F) at 4.5bya (Kasting & Ackerman 1986); conditions which would have destroyed pre-biotic molecules, proteins, naked DNA, but not microbial hyperthermophiles. And these temperatures were maintained because of volcanic activity, geothermal activity, and heat generated from bolide impacts which were nearly continuous for the first 800 million years.

Yet, during this same 800 million time of incessant extraterrestrial bombardment and despite the hellish conditions, life had already taken root on this planet, with evidence of biological activity dated to 4.28 to 4.2 billion years ago (O'Neil et al. 2008; Nemchin et al. 2008), and with fossil and additional biological evidence of life at 3.8 billion years ago (Manning et al. 2006; Mojzsis et al. 1996; Pflug 1978; Rosing, 1999, Rosing & Frei, 2004). These findings, indicating life was on Earth from the beginning and most likely arrived in extraterrestrial debris, are further supported by the analysis of molecular clocks which indicates Eubacteria and Archaebacteria were present on this planet over 4 billion years ago (Battistuzzi and Hedges, 2009).

Biological molecules would have been destroyed, but many species of microbe could easily survive these hellish conditions completely unscathed, including those who typically flourish under blistering, broiling, and boiling conditions (Boone et al. 1995; Setter 2002), as well as those which arrived safely ensconced within massive extraterrestrial debris.

The evidence, therefore, indicates that many essential ingredients necessary for fashioning proteins, nucleotides, and RNA and DNA were lacking on the early Earth, and the thermal and radioactive conditions would have destroyed any biological molecules as well as proteins and naked RNA and DNA (if by some miracle they had been fashioned). By contrast life was present on Earth from the very beginning as indicated by evidence of biology in this planet's oldest rocks. In fact, if all rocks had not been repeatedly vaporized by the incessant pounding of extraterrestrial debris, it is likely evidence of biological activity as well as microfossils would have appeared at 4.6 billion years (the supposed creation date of this planet), instead of 4.28 bya when some of the melted rocks had first begun to cool and harden.

There is no evidence life can be fashioned from non-life, at least on Earth. Therefore, the preponderance of evidence indicates life on Earth came from other planets.

Genetics Indicates a 10 Billion to 14 Billion Year Ancestry

Based on four separate analyses of the evolution of the genome, and single gene and whole genome duplicative events, and beginning with evidence of life between 4.2 and 3.5 billion years ago, Joseph and his assistant Wickramasinghe (2011) determined that beginning with a single gene, it took 10 billion to 14 billion years to fashion a minimal gene set of no fewer than 382 genes the minimum necessary to maintain the simplest of free-living bacterial life. Subsequently, Sharov and Gordon (2013) determined it would take 9.7 billion years beginning with one base pair of nucleotides, and 5 billion years to reach the complexity of a simple bacteria. Based on this data, it could be argued that the ancestry of carbon-based, DNA-based bacterial life, in this galaxy (as represented by life on Earth), extends backwards in time to at least 10 billion or more years, during a period and in locations where the chemistry and physics were ripe for fashioning those self-replicating DNA-equipped molecules whose descendants would eventually fall to Earth.

If these dates are correct, then a primitive genetic code consisting of just one gene must have first been established in an extraterrestrial environment and this led to the first extraterrestrial DNA-proto-bacterial replicon which also had to have been fashioned billions of years before Earth was formed. This DNA-replicon began to replicate, made variable copies of itself and became more complex, giving rise to a simple bacteria with a genome of several hundred genes. Through mechanisms of panspermia the descendants of these bacterial organisms were tossed from planet to planet and eventually deposited on Earth (Joseph & Schild 2010a,b).

It can also be assumed that it also took at least 10 billion years to generate a minimal gene set sufficient to sustain the life of a simple archae. It is not likely that archae evolved from bacteria or that bacteria evolved from archae for reasons which will be explained.

Genetic Origins of Life

Our Milky Way galaxy, and numerous other galaxies, are over 13 billion years old (Pace & Pasquini, 2004; Pasquini et al., 2005). The range of dates for the establishment of the first gene (10 to 14 bya) overlaps with the age of the Milky Way galaxy, and with the currently accept time frame for what many believe was the "Big Bang" beginning of this universe--though, it should be stressed that some scientists do not believe the universe was created, but is eternal and infinite with no beginning and no end. Therefore, given these two cosmological models it could be argued that life began and the first gene was established infinitely long ago in an infinite universe, or soon after the Big Bang creation event and the establishment of these galaxies including the Milky Way (Joseph & Schild 2010a).

The genetic analysis performed by Joseph and Wickramasinghe (2011) and Sharova and Gordon (2013), however, did not take into consideration the length of time necessary to form just a single protein. Hoyle (1974) calculated the probability of forming just a single protein consisting of a chain of 300 amino acids is $(1/20)^{300}$ or 1 chance in 2.04×10^{390}, and estimated it would take a trillion years and a trillion chance combinations.

In an infinite universe a trillion years or even a hundred trillion years would be more than enough time for life to have been fashioned if Hoyle's analysis is correct. The same would not be true of a created universe less than 14 billion years in age.

However, if instead of trillions of years, estimates for fashioning proteins, nucleotides, and the first strand or DNA were based on trillions of chance combinations of all the essential ingredients on the right conditions, all the components of life could have been fashioned in a finite or infinite universe or in the Milky Way Galaxy alone.

As detailed and reviewed by Joseph and Schild (2010a,b), the number of galaxies in the known, Hubble length universe might be a trillion sextillion. Each of these galaxies likely contain hundreds of billions to trillions of stars, each of which was presumably fashioned in a nebular cloud. For example, a single galaxy, such as Andromeda, may contain over a trillion stars (Mould, et al., 2008) whereas the Milky Way may have over 400 billion. Each of these stars were likely created in nebular clouds which may have contained most if not all the necessary chemicals and agents for the creation of life (Belloche, 2009; Fraser, 2002; Jura, 2005; Osterbrock & Ferland 2005; Williams, 1998; Zelic, 2002).

Therefore, given a trillion sextillion galaxies with stars and planets which are even more numerous, then chance combinations of all the necessary chemicals to form life, could have taken place in each of these star producing chemically enriched nebular clouds over billions of years of time, such that self-replicating molecules were repeatedly fashioned (Joseph & Schild 2010a). However, this does not mean that all would have achieved life. Nevertheless, given even more extreme almost improbable odds of 1 in a sextillion trillion, it can be predicted that given over a sextillion trillion environments where trillions of chance combinations took place, then life could have arisen in multiple galaxies through chance combinations of the necessary ingredients in the womb of nebular clouds.

When we consider not just nebular clouds, but the vast number of planets and comets, some of which may have also contained all the necessary ingredients for generating life, extreme odds of up to sextillion trillion are no longer daunting.

If we restrict our analysis to the Milky Way galaxy, with its 400 billion stars and its trillions of (likely) planets, and to probabilities of 1 in a trillion, then a trillion chance combinations may have occurred billions if not trillions of times somewhere in just this galaxy, until finally a self-replicating combination of molecules were fashioned (Joseph & Schild 2010a). Therefore, it could be said that life may have have begun by 10 to 14 billion years ago, in this galaxy, perhaps right around the time or soon after this galaxy was formed 13 billion years ago.

On the other hand, although it takes at least 10 billion years to go from a single gene to a minimal gene set necessary for life, this does not mean that life began 10 billion years ago, but only that it takes at least 10 billion years. Again, the genetic analyses performed by Joseph, Sharov and Gordon did not include estimates as to the formation of the first proteins and nucleotides or the creation of the first DNA macro-molecule; and this leads us back to those estimates ranging from a trillion years to completely improbable. However, if the universe is infinite the first gene and the first minimal gene set could have been established infinitely long ago.

Life may have had no beginning, or the universe itself may be alive. Or, as the atomists posited over two thousands years ago, within an infinite universe the atoms of life may continually assemble and reassemble over infinite time, such that, in the final analysis, life continually recycles itself, which means, life comes from life.

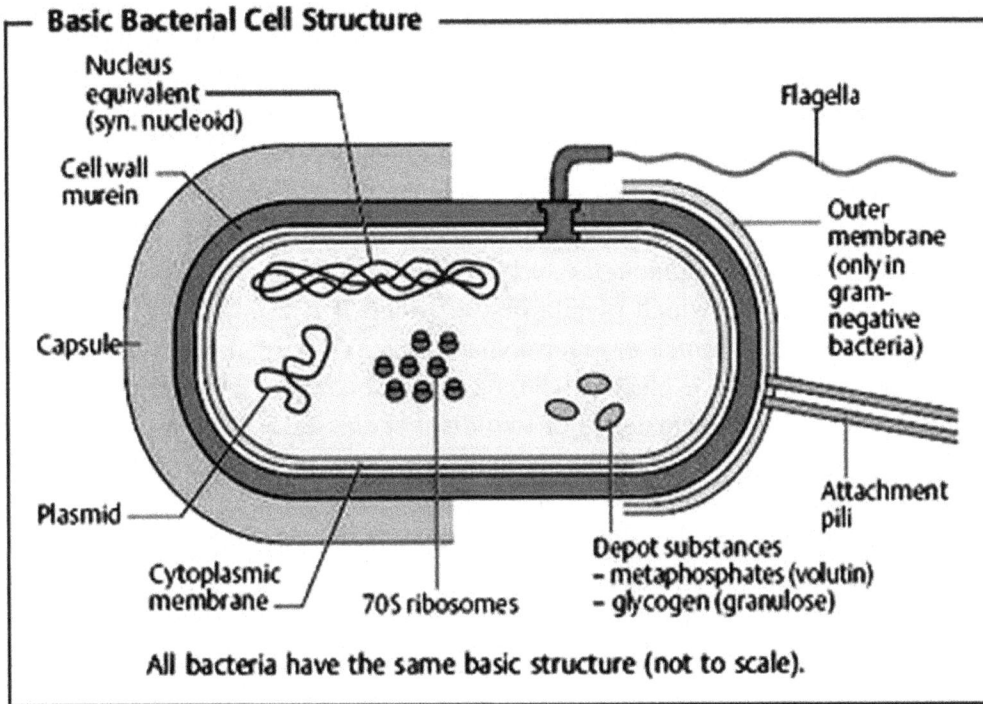

Basic Bacterial Cell Structure

Nucleus equivalent (syn. nucleoid)

Flagella

Cell wall murein

Outer membrane (only in gram-negative bacteria)

Capsule

Plasmid

Attachment pili

Cytoplasmic membrane

70S ribosomes

Depot substances
- metaphosphates (volutin)
- glycogen (granulose)

All bacteria have the same basic structure (not to scale).

The DNA Cosmic Imperative: Universality of the DNA Genome

That the three domains of life (archae, bacteria, and eukaryotes) all possess a DNA-based genome, whereas viruses have an RNA or DNA genome, could be considered evidence for common origins from a single source. Thus, after a trillion chance combinations in an ideal extraterrestrial environment, life was fashioned, and then diverged, and diverged again, giving rise to archae, bacteria, viruses, and Eukaryotes. However, although there is substantial evidence that the co-mingling of Prokaryote and viral genes contributed to what became the multi-cellular Eukaryotic genome, there is no evidence that bacteria can become an archae or an archae a bacteria, or that a bacteria or a archae can become a Eukaryote. Rather, it took a commingling of genes from different species to form a multicellular Eukaryote. If life only began once in the vastness of the cosmos, how could it split up into three domains of life, two of which (archae and bacteria) are distinctly different, generally do not associate together, and often dwell in completely distinct environments?

Archae and bacteria are considered Prokaryotes. In the broadest terms, archaea are distinct from bacteria, particularly in regard to the size of their genomes and cell membranes. For example, archaean membranes are made of ether lipids whereas bacterial cell membranes are created from phosphoglycerides with ester bonds (De Rosa et al., 1986). Like bacteria, archae can live in the most extreme environments (Kimura et al, 2006, 2007; Leininger et al., 2006; Robertson et al., 2005). However, whereas bacteria are usually the most common form of life in the soil, archae are the most common form of life in the ocean, dominating ecosystems below 150 m in depth (Karner et al., 2001; Robertson et al., 2005).

The genomes of archae are rather uniform and compact in size ranging from 0.5 Mb in the parasite *Nanoarchaeum equitans* (Waters et al., 2003) to 5.5 Mb in *Methanosarcina barkeri* (Maeder et al., 2006).

Bacterial genomes can vary by two orders of magnitudes, from 180 kb in the intracellular symbiont, *Carsonella rudii* (Nakabachi et al., 2006), to 13 Mb in *Sorangium cellulosum* which dwells in soil (Schneiker et al., 2007). Although there are bacterial genomes of intermediate size, the vast majority of bacteria so far sequenced show a clear-cut bimodal distribution of genomes; i.e. large vs small, suggesting the existence of two distinct classes of bacteria: those with 'small' genomes (Ranea et al., 2005) with the highest peak at 2 Mb and those with "large" genomes at about 5 Mb (Schulz & Jorgensen 2001).

Although speculation abounds, there is no convincing evidence that archae and bacteria originated from a common ancestor which, in fact, has never been found. On the other hand, if that common ancestor dwells only in specific extraterrestrial environments, then perhaps the common ancestor to all of life will someday be discovered.

However, if life began multiple times or in multiple extraterrestrial locations, and if archae, bacteria, Eukaryotes and viruses have separate or even overlapping origins, why would the genetic code be similar, almost universal among all species, all domains of life, as well as viruses?

The universality of the DNA-genome and the genetic code, may indicate that DNA is a "cosmic imperative" and a requirement for life, and everything else and all other codes fail to achieve life. Thus, where ever life is fashioned, be it in a primordial planet, a nebular cloud in another galaxy, a comet traveling through interstellar space, pre-life can only achieve life if it contains DNA and a genome with a minimal gene set.

Be it a finite created Big Bang universe, or an infinite universe with infinite time, then through "natural selection" (as conceived by A. R. Wallace) coupled with endocytosis, phagocytosis, and horizontal gene transfer, that over billions of years extreme variations in genetic coding between innumerable microbial species were eventually *averaged* out; or that one genetic code won out as it was the superior code.

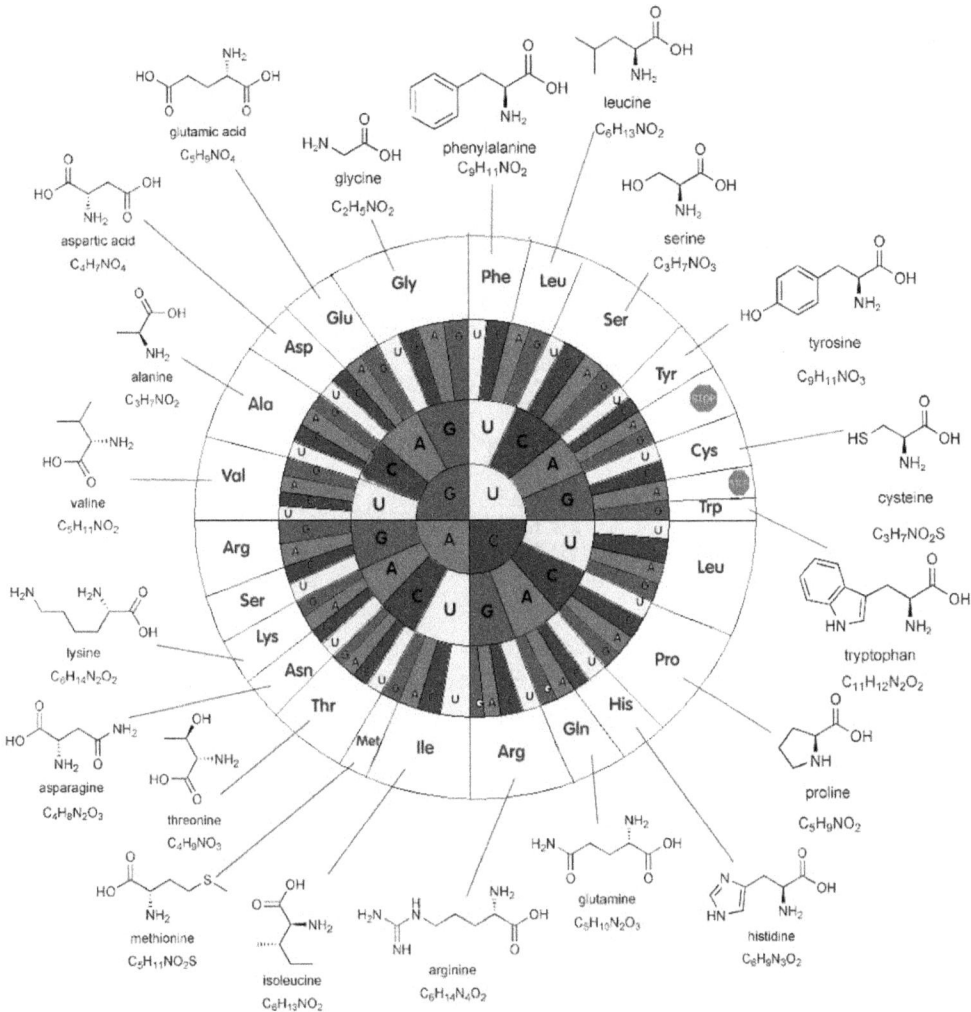

glutamic acid
$C_5H_9NO_4$

leucine
$C_6H_{13}NO_2$

phenylalanine
$C_9H_{11}NO_2$

glycine
$C_2H_5NO_2$

serine
$C_3H_7NO_3$

aspartic acid
$C_4H_7NO_4$

alanine
$C_3H_7NO_2$

tyrosine
$C_9H_{11}NO_3$

valine
$C_5H_{11}NO_2$

cysteine
$C_3H_7NO_2S$

lysine
$C_6H_{14}N_2O_2$

tryptophan
$C_{11}H_{12}N_2O_2$

asparagine
$C_4H_8N_2O_3$

threonine
$C_4H_9NO_3$

proline
$C_5H_9NO_2$

methionine
$C_5H_{11}NO_2S$

glutamine
$C_5H_{10}N_2O_3$

histidine
$C_6H_9N_3O_2$

isoleucine
$C_6H_{13}NO_2$

arginine
$C_6H_{14}N_4O_2$

Thus, initially, even if completely diverse life forms were generated in different extraterrestrial environments and with wildly different genetic codes, those codes may have become modified or extinguished upon encountering DNA-based life with a superior genetic code. Through well established mechanisms of horizontal gene transfer (Polz, et al., 2013), the genes based on the superior genetic code were inserted into genomes without universal codes (Joseph & Schiled 2010a,b). The superior code and its superior genes and genome then began dominating and taking over inferior genomes, thereby giving rise to a universal genetic code which is common to all life.

Bacteria   **Archaea**   **Eucarya**

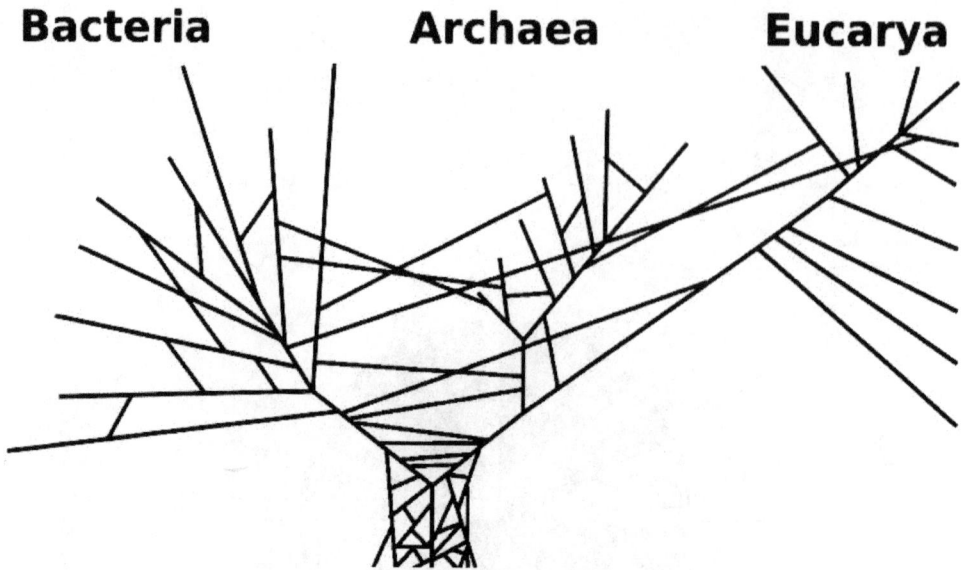

If these latter propositions are true, then the different domains of life could have arisen in completely different extraterrestrial environments under localized conditions where all the essential ingredients were available for the manufacture of proteins, enzymes, DNA, RNA, the cell wall, membranes, etc. Thus, a bacteria (or archae) may have been generated in nebular clouds, an archae (or bacteria) within the interior of comets, and a singled celled Eukaryote (or an archae or bacteria) within the oceans of an ancient planet billions of years before Earth became Earth (Joseph & Schild 2010a,b). And in each instance their origins may have been preceded or accompanied by viruses which could have acted as "mobile RNA worlds" (Joseph 2000a). Later, upon making contact, bacteria, archae, Eukaryotes, and viruses exchanged genes which resulted in a universal genetic code.

And then, via mechanisms of panspermia (Joseph & Schild 2010b) the descendants of these archae, bacteria, and viruses, were repeatedly deposited on and then later ejected from innumerable planets. And with each expulsion and then crash landing onto the surface of yet other worlds, these creatures engaged in horizontal gene transfers with each other and all species encountered, until finally after billions if not trillions of years some of these "genetic seeds of life," fell to Earth.

And this is how life on Earth began. And then, it began to evolve.

3. Cambrian Explosion: Silent Genes, HGT, Cyanobacteria, Bones, Brains, Eukaryotic Metamorphosis, Genetic Engineering of the Environment

Life Was Present on Earth From The Beginning: Viruses, Archae, Bacteria

Most scientists believe the genetic ancestry of modern day life can be traced to the first living creatures to appear on Earth. However, life was present on Earth from the very beginning, as demonstrated by discoveries of biological activity in the first rocks to re-solidify and harden, dated to 4.28 billion years (O'Neil et al, 2008) and 4.2 bya (Nemchin et al. 2008). Additional evidence of life has been dated to 3.8 bya, and this includes carbon-isotopes discovered in quartz-pyroxene rocks on Akilia, West Greenland, and within a phosphate mineral, apatite, which includes tiny grains of calcium and high levels of organic carbon--the residue of photosynthesis, oxygen secretion, and thus biological activity (Manning et al. 2006; Mojzsis et al. 1996). And this biological activity was most likely that of Cyanobacteria which is the only known species of Prokaryote capable of photosynthesis, and which also secretes oxygen and calcium.

In addition, microfossils resembling Eukaryotic yeast cells and fungi were discovered in 3.8 billion year old quartz, recovered from Isua, S. W. Greenland (Pflug 1978). Evidence of biological activity including photosynthesis was also discovered in this area dated from the same time period (Rosing 1999, Rosing & Frei 2004).

It was during this same time period, 4.28 to 3.8 bya, that Earth was incessantly bombarded by extraterrestrial debris (Belbruno & Gott, 2005; Rankenburg et al. 2006; Schoenberg et al. 2002) creating thermal and other disturbances which would have destroyed all biological and prebiotic molecules (Crick 1981; Ehrenfreund & Sephton 2006) while leaving unscathed at least some of those microbes buried within that extraterrestrial debris, or those who may have been dwelling deep beneath the earth. It can be deduced, therefore, that the first Earthlings arrived here within those meteors, asteroids and comets, and that Cyanobacteria were among these sojourners from the stars.

It can also be assumed that these first Earthly creatures included archae, bacteria (Battistuzzi & Hedges, 2009), and by 3.8 bya single celled Eukaryotes (Pflug 1978), all of which would have been accompanied by viruses (Joseph 2000a, 2009b; Joseph & Schild 2010a,b). And it can be concluded that the ancestors of these microbes and viruses hailed from space and with a genetic ancestry leading to planets much older than Earth.

Viruses are found in association with and outnumber archae and bacteria on ratios ranging from 1 to 10, and 1 to 100 respectively. Although viruses are associated with disease, collectively they serve as vast storehouses of genes and gene depositories (Joseph 2009b,c,d) which can be transferred to archae and bacteria (as well as to Eukaryotes) on "as needed basis" (Lindell et al., 2004; López-Sánchez et al., 2005; Sullivan et al., 2006; Romano et al., 2007). In fact, viruses can store multiple genomes, with 25% of viruses so far studied encapsulating up to three complex genomes (Goff et al., 2012). Thus, when viruses as well as Prokaryotes arrived on this planet in successive waves of invasion from space, they carried with them genes, and possibly entire genomes which had been acquired via horizontal gene transfer, from living creatures which had evolved on other worlds (Joseph 2000a, 2009b,c,d). And once on Earth, these patterns of horizontal gene transfer continued, resulting in a co-mingling of genes which resulted in the generation of the first Earthly multi-cellular Eukaryotes.

Almost from the very beginning of the establishment of life on Earth, Prokaryotic (archae and bacteria) genes and viral genes were contributed to the Eukaryotic genome; and these donated genes have influenced and shaped the evolution of all subsequent multi-cellular species leading to humans. They also terraformed the planet in preparation for those yet to evolve, and in so doing liberated minerals and released gases such as oxygen, which acted on those genes they had inserted into Eukaryotes.

Prokaryotes (Cyanobacteria in particular) have biologically engineered this planet such as by pumping oxygen and calcium into the air and sea. Earth's changing environment in turn acted on gene selection--genes which were transferred to the Eukaryotic genome by Prokaryotes and viruses and which were obtained via horizontal gene transfer from species who evolved on other worlds (Joseph 2000a, 2008, 2009b,c,d). And these genes, once embedded in the genomes of Eukaryotes, released pre-coded traits in response to the biologically engineered changing environment which enabled oxygen breathing animals to evolve hearts, bones and brains, then crawl upon the surface of the planet, walk on four then two legs, then stand upright and gaze into the heavens to ponder the nature of existence.

What has taken place on Earth is not a Darwinian evolution, but has unfolded according to the basic principles governing metamorphosis and embryogenesis and which resembles the precise, purposeful and genetically regulated interactions of those cells comprising a complex multi-cellular supra-organism.

Silent Genes, Horizontal Gene Transfer, Evolution

Genomic analysis has demonstrated that genes are commonly shared between bacteria and archaea (Koonin 2009a,b; Polz et al 2013), between Prokaryotes and Eukaryotes (Hotopp et al., 2007; Martin et., al., 2002; Nikoh et al., 2008), between viruses and Prokaryotes and other viruses (Lindell et al., 2004; Sullivan et al., 2006), viruses and Eukaryotes (Conley et al., 1998; López-Sánchez et al., 2005; Romano et al., 2007), and between Eukaryotes (Maruyama et al. 2011). In fact, genes and entire chromosomes can be transferred between Eukaryotes including between humans (Berger et al., 2013; Chen et al. 2013; Dhimolea et al., 2013; Lupski, 2013). The sharing of genes is accomplished via horizontal gene transfer (HGT).

For example, in one study, 56% of healthy women were found to have the male Y chromosome in their brest tissue and that male cells had also invaded these tissues (Dhimolea et al., 2013). Although genetic mosaicism and chimerism are often associated with disease (Biesecker & Spinner 2013; Lupski, 2013), Dhimolea and colleagues (2013) found that these genes provided a protective function. Berger and colleagues (2013) also discovered a mixture of genomes (the donors and the recipients) in 74% of those individuals who had undergone a medical transplant and in whom the operation was successful.

However, although the transfer of genes between Eukaryotic genomes can provide benefits or protection (Dhimolea et al., 2013), generally when Eukaryotes (including humans) transfer genes to one another this typically creates chimeras and genomic mosaics, and the result is mutation and disease (Biesecker & Spinner 2013; Lupski, 2013; Poduri et al., 2013). Mutations do not contribute to the evolution of species.

By contrast, genes transferred to the Eukaryotic genome by Prokaryotes and viruses, provide substantial benefits to the host and its genome. Transferred genes and genetic material include exons, introns, transposable elements, informational and operational genes, RNA, ribosomes, mitochondria, and the core genetic machinery for translating, expressing, and repeatedly duplicating genes and the entire genome (Charlebois & Doolittle 2004; Dehal & Boore 2005; Koonin & Wolf 2008; Lynch et al. 2001). Archae, bacteria, and viruses, provided Eukaryotes with the regulatory elements which control gene expression and which have repeatedly duplicated individual genes and the entire genome thereby enabling the Eukaryote gene pool to grow in size and leading to evolutionary innovation and the generation of increasingly intelligent species.

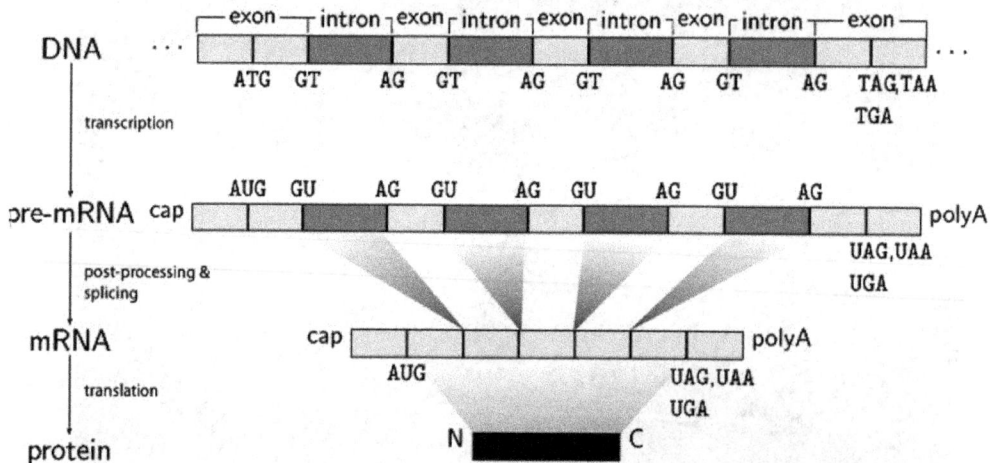

Even introns, ribosomal proteins, and RNA polymerase subunits are subject to HGT (Brochier et al., 2000; Iyer et al., 2004) and the same is true of regulatory genes, many of which have been inserted into the Prokaryotic and Eukaryotic genome by viruses. For example, group II introns and ribozymes which are derived from viruses are found in the genomes of archae, bacteria, and Eukaryotes (Dai &

Zimmerly 2003; Dai et al., 2003).

Thus we see that the genomes of modern day Eukaryotic species, including humans, contain highly highly conserved genes which were acquired from archae, bacteria, and viruses (Esser et al. 2007; López-Sánchez et al., 2005; Rivera & Lake 2004; Romano et al., 2007; Yutin et al. 2008).

Viruses, as well as bacteria and archae, can also store their genes within the Eukaryotic genome (Conley et al., 1998; López-Sánchez et al., 2005; Romano et al., 2007). In fact, 8% of the human genome consists of around 200,000 endogenous retroviruses (IHGSC 2001; Medstrand et al., 2002), and 3 million retro elements (Medstrand et al., 2002), and some of these retroviruses are still active whereas others are silent and have yet to be expressed (Conley et al., 1998; Medstrand & Mager, 1998).

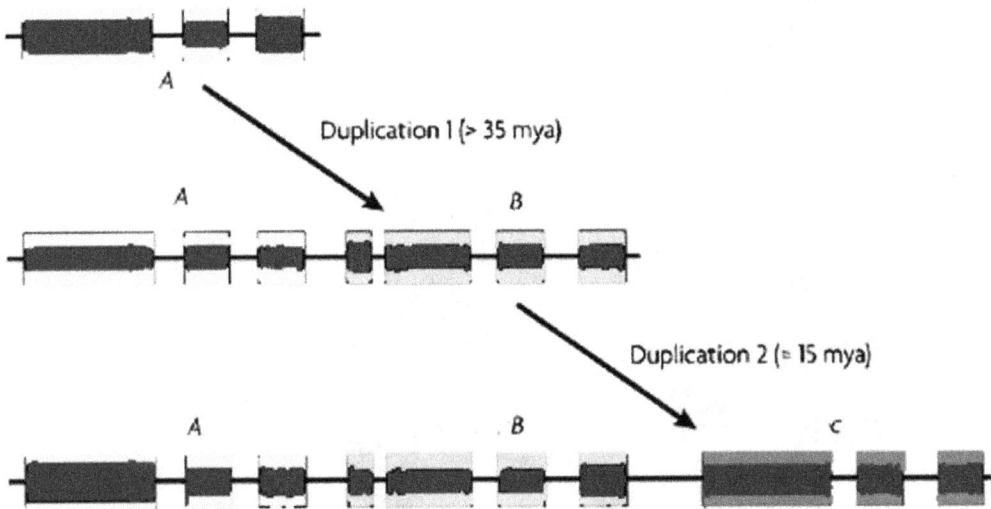

The First Eukaryotes

The first multi-cellular Eukaryote may have been fashioned when certain archae, bacteria, and viruses exchanged and shared genes and formed symbiotic/parasitic relationships, thus creating a combined genome within a single cell--a process reminiscent of sperm-ovum fertilization and embryogenesis. The combination of these Prokaryote and viral genes therefore produced a new genome and a new species and possibly a new domain of life: Eukaryotes (Feng et al., 1997; Joseph 2009b, 2010a; Martin & Muller, 1998; Rivera & Lake 2004). However if these combined genes generated the first unicellular Eukaryotes (Woese 2004) or if single celled Eukaryotes were transformed into a multi-cellular Eukaryote after single celled Eukaryotes were first generated in their own unique extraterrestrial environment, is unknown.

Following successive invasions and combinations of Prokaryote and viral genes, some species of unicellular Eukaryotes became more diverse, and some of their descendants later came to be comprised of compartments and a nucleus which contains the DNA of multicellular Eukaryotes. However, the nucleus and the other compartments may have originally consisted of symbiotic or invading archae and bacteria which were engulfed and phagocytized and stripped of their genes (Lake & Rivera 1994; Horiike et al. 2004). Likewise, organelles, as well as mitochondria, may have been created following engulfment and the donation of bacterial and archae genes to the Eukaryotic host (Dyall et al., 2004; Embley & Martin, 2006; Margulis et al., 1997; Pace 2006; Woese 1994). The incorporation of these Prokaryotic genes and the symbiotic relations which developed between Eukaryotes and genetically-stripped down bacteria and archae, led to the creation of the nucleus and compartmentalization.

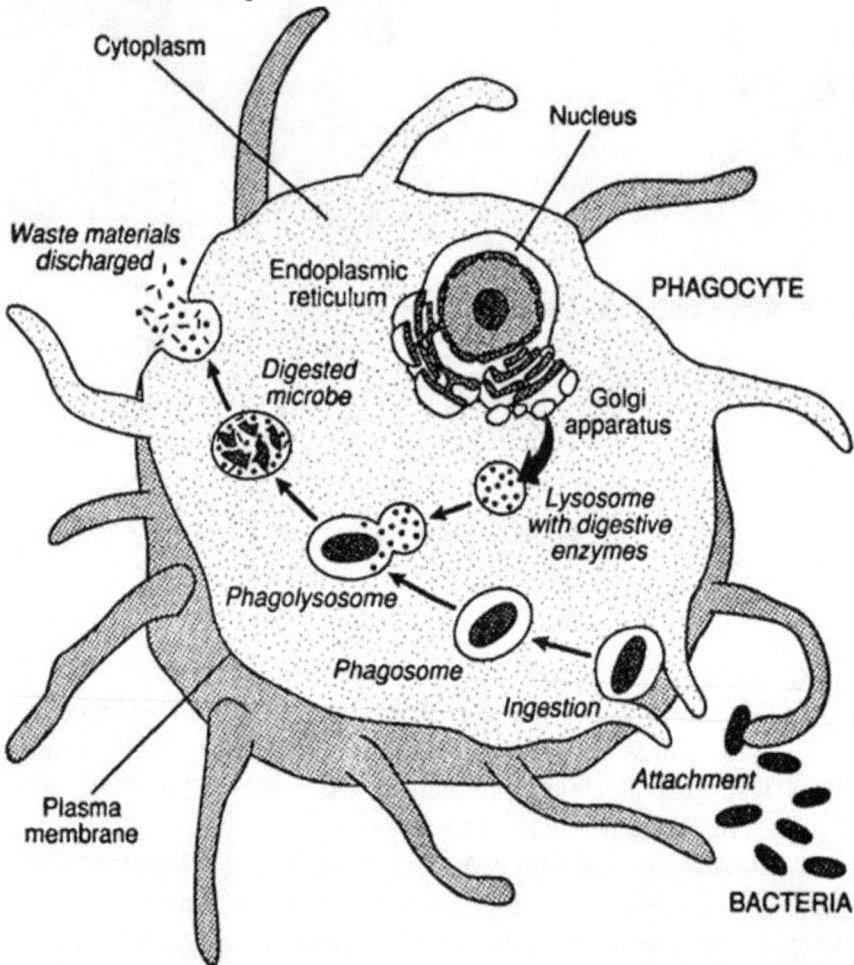

The nucleus and compartmentalization made it possible for predatory Eukaryotes to ingest and phagocytize other creatures while minimizing the risk of random gene mixing and the unregulated incorporation of foreign DNA. Therefore, it appears the Eukaryotic nucleus was fashioned out of a Prokaryote which provided genomic protection, and this allowed other microbes to be safely ingested or incorporated thereby giving rise to additional compartments including the metamorphosis of mitochondria.

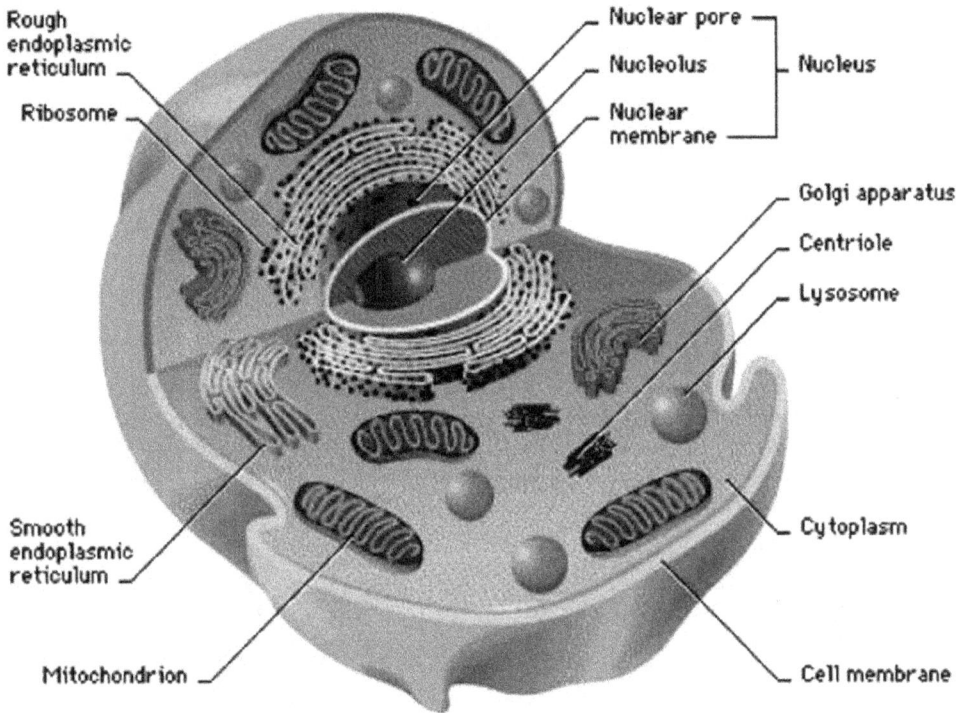

However, as the same time, these stripped down internalized Prokaryotes also allowed specific Prokaryote genes including viral genes and elements to be incorporated into the Eukaryotic genome following HGT. Therefore, Prokaryotes not only provided genes, compartments, and nuclei, but acted as gatekeepers which could determine which genes would be accepted and which would be rejected.

These developments enabled Eukaryotes to become more complex and conquer new environments which then acted on gene selection.

When some single celled Eukaryotes acquired additional symbiotic partners and genes donated by archae, bacteria, and viruses, they became multi-cellular (Joseph 2009d, 2010a). Therefore, the first multi-cellular Eukaryotes were fashioned via the combination of Prokaryotic and viral genes; an event which may commonly take place on every habitable planet.

Thus we see that the genomes of modern day Eukaryotic species, including humans, contain compartments, nuclei, mitochondria which appear to be stripped down Prokaryotes, and they contain highly conserved genes which were acquired from archae, bacteria, and viruses (Esser et al. 2004, 2007; Rivera & Lake 2004;

Yutin et al. 2008). However, not all of these genes have been expressed, where-as yet others have been activated in response to specific environmental signals, thereby giving rise to new species (Joseph 2000a, 2009b,c).

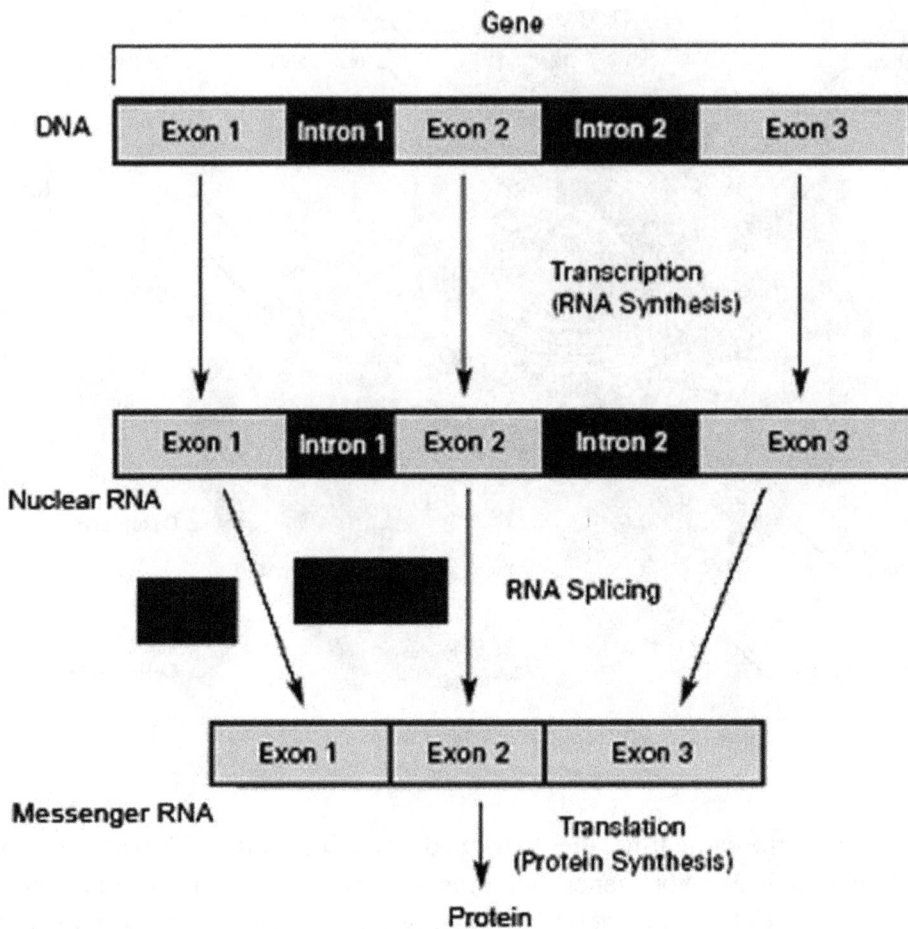

Genes transferred to the Eukaryotic genome by Prokaryotes and viruses, include exons, introns, transposable elements, informational and operational genes, as well as RNA, ribosomes, mitochondria, and the core genetic machinery for translating, expressing, and repeatedly duplicating genes and the entire genome (Charlebois & Doolittle 2004; Dehal & Boore 2005; Harris et al. 2003; Koonin & Wolf 2008; Lynch et al. 2001; McLysaght et al. 2002). Archae, bacteria, and viruses, provided Eukaryotes with genes which code for core cellular functions including the regulatory elements which control gene expression and which have repeatedly duplicated individual genes and the entire genome. Individual and whole genome duplications have enabled the Eukaryote gene pool to grow in size thereby leading to evolutionary metamorphosis and the generation of increasingly complexand intelligent species, including humans.

Moreover, these donated Prokaryotic genes, including a core set of approximately 70 genes, have been conserved and passed down, without deletion, for billions of years, and make up between 1% to 10% of the genes in the genomes of all multicellular life including humans (Koonin & Wolf, 2008; Harris et al., 2003; Charlebois & Doolittle 2004).

These conserved genes, proteins, and gene sequences, include those governing translation, the core transcription systems, and several central metabolic pathways, such as those for purine and pyrimidine nucleotide biosynthesis (Koonin 2002, 2003, 2009b). Moreover, protein sequence conservation extends from mammals to bacteria thus demonstrating their great antiquity (Dayhoff et al., 1974, 1983).

Archae, Bacteria and Operational and Informational Genes

Broadly considered, the Eukaryote genome contains two sets of functionally distinct prokaryotic genes, operational vs informational; one set derived from archaea and the other from bacteria.

Archae provided the Eukaryote genome with genes for information processing and expression (translation, transcription, replication, and repair). Over 350 eukaryotic genes have been identified that are of apparent archaeal origin and which were acquired via early horizontal gene transfer (Yutin et al. 2008). In fact, an analysis of ribosomal structure and ribosomal protein sequences indicates a specific affinity between Eukaryotic genes and their orthologs from archae (Lake 1988; 1998; Rivera & Lake 2004; Vishwanath et al. 2004). Thus, archae were a major source of introns and transposable elements.

By contrast, bacteria provided operational genes responsible for the Eukaryotic membrane system, the inner cytoskeleton, complex metabolic activity, metabolic enzymes, and the production of the principal enzymes of membrane biogenesis (Esser et al. 2004, 2007; Pereto et al. 2004; Rivera & Lake 2004; Yutin et al. 2008). These include genes and proteins which directly influence metabolism and the ingestion and excretion of various waste products. Operational genes have been repeatedly and continuously horizontally transferred over the course of evolution (Jain et al. 1999).

The combination of these two sets of genes, informational and operational, which were donated by Prokaryotes to the Eukaryotic genome, contributed significantly to the evolution of Eukaryotic complexity. Likewise, many of the proteins that regulate Eukaryotic signal transduction networks, including those involved in apoptosis and programmed cell death, are derived from the Prokaryotic genome (Aravind et al. 1999; Bidle & Falkowski 2004; Koonin & Aravind 2002). These signaling molecules are common in bacteria, Cyanobacteria, and archae and include proteases from the AP-ATPase family. These proteases perform catalytic functions, and are found in the plant and animal genome (Bidle & Falkowski 2004; Koonin & Aravind 2002) and are utilized by mitochondria. Therefore, the genes coding for these products were also transferred from Prokaryotes to Eukaryotes.

The fashioning of multicellular Eukaryotes and the contribution of so many crucial genes, was no random act of nature, but was purposefully orchestrated and choreographed, much like embryogenesis and metamorphosis which are precisely genetically regulated. These complex interactions should not be viewed as independent acts by cells acting as lone wolves, but the complex acts of a supraorganism consisting of genes linking numerous species. Thus we see that various species of Prokaryote donated genes whereas other Prokaryotes invaded and shed genes and in so doing acted as compartmentalized and nucleated gatekeepers which determined if and what "foreign" genes could be introduced into the Eukaryotic genome. Genes which were allowed entry included regulatory genes and elements contributed by viruses and Prokaryotes who in turn biologically engineered the environment which led to the expression of those genes.

These interactional patterns are not randomized acts of individual cells, but genetically regulated and purposeful, much like the inner compartments of a single cell act to maintain the functional integrity and growth of that cell; or the various cells of a single organism maintain interdependent relationships which promote the health and development of that organism. In other words, cellular *individuality* is an illusion, and that includes the seemingly independent actions of Prokaryotes and viruses which are not behaving in isolation but purposefully in relation to numerous microbial species so as to promote the development and metamorphosis of Eukaryotes with whom they are genetically entwined and linked.

Mitochondria Metamorphosis: Bacteria Invade, Engineer Eukaryotes

Almost two billion years after the first unicellular Eukaryotes appeared on the surface of this planet, a mitochondria-like bacteria may have donated its genome after invading a Eukaryote (Margulis et al., 1997). The internalized bacteria and Eukaryote may have formed a mutual genetic as well as a symbiotic relationship which resulted in the metamorphosis of mitochondria. These mitochondria enabled Eukaryotes to breath the oxygen being secreted by Cyanobacteria, and greatly increased the cell's capacity to extract and generated energy.

Prior to this epic event, Eukaryotes may have consisted of less than 2 cell types (Hedges et al., 2004). However once the bacterial invasion and its transformation into a mitochondria was complete, Eukaryotes soon grew in size and complexity.

Mitochondria live inside every single cell of every multi-cellular organism, adjacent to the nucleus. The genomes of all extant multi-cellular creatures contain genes which can be traced to ancestors that possessed a bacterial endosymbiont that gave rise to the mitochondria (van der Giezen & Tovar 2005; Embley 2006).

Many cells have only a single mitochondrion, whereas others contain several thousand. Mitochondria have their own independent genomes and their DNA shows substantial similarity to bacterial genomes (Pace 2006; Woese 1994). Mitochondria are enclosed in their own inner and outer membrane, play a significant role in signaling, cellular differentiation, cell death, as well as the control of the cell cycle and cell growth (Chipuk et al., 2006; Mannella 2006; Rappaport et al., 1998).

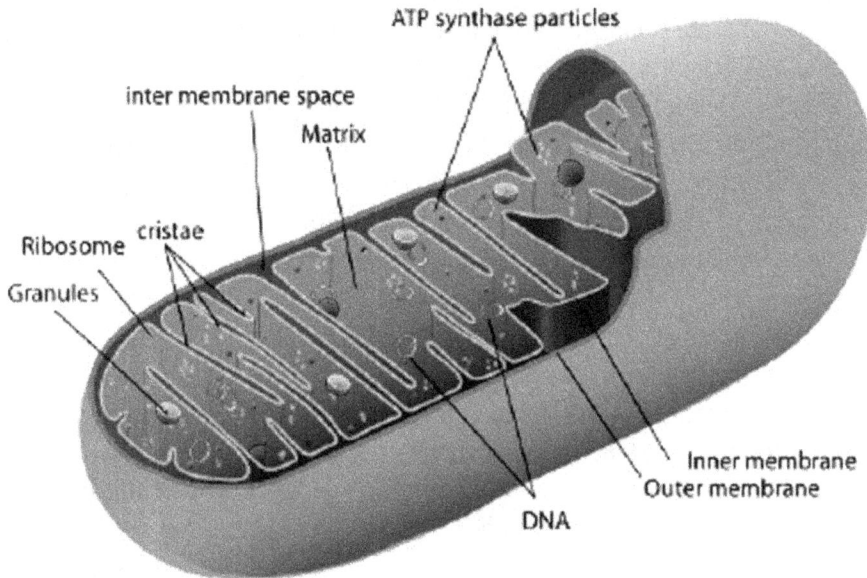

Mitochondria also serve as the powerhouse of the cell and are located outside the nucleus. Mitochondria generate most of the cell's supply of adenosine tri-phosphate (ATP) which is used as a source of chemical energy. The production of ATP is accomplished by oxidizing the major products of glucose, pyruvate, and NADH, which are produced in the cytosol (Akao et al., 2001; Dahout-Gonzalez et al., 2006; Garlid et al., 2003).

Thus, mitochondria are essential to the functioning of the cell, providing these organisms with substantial energy, and enabled Eukaryotes to grow larger in size and exploit the changing biologically engineered environment, which in turn acted on gene selection.

This invasion and gene transfer which resulted in the first mitochondria began around 2.3 bya during a time when the environment was beginning to become oxygenated (Barleya et al., 2005; Mentel & Martin 2008) and enriched with sulphide and ferrous iron which served as oxygen acceptors (Sleep & Bird 2008).

As oxygen levels increased, methane levels decreased, and the Earth became glaciated, fueled by oxygenic photosynthesis (Joseph 2009d, 2010a). This rise in oxygen has been referred to as the Paleoproterozoic "Great Oxidation Event" (~2.2 to 2.0 Ga), when atmospheric oxygen may have risen to >1% of modern levels, a byproduct of oxygenic photosynthesis by Cyanobacteria (Buick 2008; Canfield 2005; Holland 2006; Nisbett & Nisbet 2008; Olson 2006).

The spike in oxygen and the onset of this period of biologically induced glaciation acted on gene selection. The internalized bacteria underwent metamorphosis and became a mitochondria and oxygen-dependent ATP-generating pathways replaced the less efficient oxygen-independent pathways and Eukaryotic cells underwent a significant alteration and began breathing oxygen. The activation of the genes donated by Prokaryotes, and the metamorphosis of mitochondria enabled Eukaryotes to colonize emerging oxygenated environments; with the oxygen being produced biologically. Mitochondria, as a distinct entity within Eukaryotic cells, were fully established by 1.8 BYA (Mentel & Martin 2008).

Thus, a complex interaction involving Cyanobacteria and oxygen production, and genes contributed by an invading bacteria (as well as other Prokaryotes including Cyanobacteria), produced an internal structure within Eukaryotes which would give Eukaryotes the ability to breath oxygen and become more powerful. It can be predicted that identical interactions and events have taken place on every habitable planet infested with these "genetic seeds of life."

The compartmentalization and nucleation of Eukaryotes and the metamorphosis of mitochondria and thus the evolution from single celled Eukaryote to multicellular Eukaryote did not take place as random acts of nature or according to Darwin's "small steps." Rather, these were precisely regulated well choreographed interactions, similar in some respects as to what take place during embryological development, such as when embryonic cells migrate, interact and establish complex structures such as the brain (Joseph 2000a,b)

Because of continued genetic interactions between Prokaryotes and Eukaryotes multicellular Eukaryotes became equipped with additional viral and Prokaryotic genes (Alvarez-Ponce et al. 2013; Forterre & Prangishvili 2013), and with compartments, mitochondria, and a nucleus consisting of stripped down Prokaryotes which were acting as genetic gatekeepers. Then as Prokaryotes, Cyanobacteria in particular, biologically engineered the external environment by releasing oxygen and liberating other elements and minerals, the changing environment acted on genes donated to Eukaryotes by viruses and Prokaryotes and the the next stage of evolutionary metamorphosis began to unfold. Eukaryotes diversified into numerous microscopic species. However, hidden away within their genomes were thousands of silent genes which would not be expressed for another billion years.

Silent Genes

The primary genetic sources for genes within the Eukaryotic genome are Prokaryotes and viruses (Alvarez-Ponce et al. 2013; Forterre & Prangishvili 2013). Be they derived from archae, bacteria, or viruses, not all these genes have been activated or expressed. Many are silent or have been silenced (Appasania, 2012; Guetg, et al. 2012; Perriaud et al., 2012; Poleshko et al. 2012).

Genes inserted into the Eukaryotic genome may be stored for hundreds of millions and even billions of years, passed on to offspring and subsequent species via the germ line. If these genes were inserted 4 billion years ago, or ten thousand years ago, once incorporated into the host genome they can be transmitted in "silent" non-activated form to daughter cells, and to subsequent generations and species for hundreds, thousands, millions or billions of years, only to be expressed in response to specific genetic regulatory and environmental signals (Ackermann et al., 1987; Appasania, 2012; Brussow et al., 2004; Joseph 2000a), thereby giving rise to increasingly intelligent species.

Silent Genes & Evolution: Bones and Brains

"Placozoa" (Trichoplax) and the Silicarea sponge have no eyes, no heart, and no brain, skeletal, or nervous system tissue, and evolved around 635 million years ago. And yet, although they apparently diverged from a common ancestor over a billion years ago, their genomes contained the same exact (silent) genes necessary for generating hearts, bones, a brain and nervous system (Sakarya et al., 2007; Srivastava et al., 2008). And yet, these genes remained silent. These silent genes inherited by Trichoplax and the Silicarea sponge were then passed on to subsequent species at which point, around 540 million years ago, these genes became activated--after 100 million years had elapsed.

How did different brainless heartless boneless species who diverged from a common ancestor perhaps over 1 billion years ago, somehow randomly "evolve" in parallel, the same genes responsible for the nervous and skeletal system; and why did these genes remain "silent" and in a non-activated state until 540 million years ago? Darwinian apologists claim this is just nature arriving, by chance, at the same solution. A solution to what? These genes were inherited from even more ancient ancestral species which also never evolved a heart or bones or a nervous system; and the ancestors of these ancestral species obtained many of their genes and genetic elements from Prokaryotes and viruses almost 4 billion years ago (Joseph 2009b,c,d).

After billions of years these silent genes and the genetic elements for generating these genes, were finally inherited by Trichoplax and other species including the sponge 635 millions years ago, where they remained dormant and silent. These silent genes were then passed down for another hundred million years through subsequent generations and species and then became activated in response to significant alterations in the environment such as an oxygen atmosphere, the produciton of ozone, and copious amounts of calcium which were flooding the oceans, thereby giving rise to the heart, eyes, bones and brains in hundreds of diverse species at the onset of the Cambrian explosion 540 mya (Joseph 2000a, 2009c,d).

These genes did not randomly evolve through natural section, for if that were true, why would they evolve and then remain silent after the evolution of Trichoplax Placozoans? Why would Trichoplax Placozoans evolve the genes for generating hearts, eyes, bones and brains, but then fail to evolve these structures? Darwinism is not the answer.

The same questions could be asked about tadpoles which possess the silent genes which when activated produce a frog, or a caterpillar which posses the silent genes which generate a butterfly. Silent genes are expressed only in response to specific regulatory and environmental signals, thereby transforming a tapole into a frog and a caterpillar into a butterfly. The same principles of metamorphosis can be applied to the *evolution* of bones and brains.

It was only after major changes in the environment, engineered primarily by Cyanobacteria, which resulted in the activation of these silent genes, and only after another 100 million years had passed. These environmental changes included a UV radiation blocking ozone layer created by the release of oxygen generated by Cyanobacteria, and the flooding of the oceans with calcium, also produced by Cyanobacteria (Joseph 2009c,d). It was a confluence of genetic, environmental and other factors, which triggered the expression of those genes responsible for hearts, eyes, a skeletal system and brain, and made it possible for species to emerge from the sea and take to the land. These were not random acts of chance, but under precise genetic control similar to what takes place during embryogenesis and metamorphosis; and exactly what might be expected of numerous species acting in concert as if part of the same supra-organism.

Genetic Engineering of the Environment: Oxygen and Cyanobacteria

Cyanobacteria are the only known Prokaryotes capable of oxygenic photosynthesis (DesMarais 2000). Oxygen is secreted as a waste product by Cyanobacteria, a species which was most likely among the first to arrive on Earth. In fact, fossilized colonies of Cyanobacteria have been discovered in the Murchison meteorite (Hoover, 2011) which is older than this solar system (Joseph 2009a).

By 3.5 bya, Cyanobacteria had begun to proliferate on the surface, protected from deadly UV rays by green house gasses. By 3.46 bya these photosynthesizing microbes had released significant amounts of oxygen into the atmosphere and oceans (Hoashi et al., 2009). In fact, they were performing the same functions from deep beneath the sea and were congregating near undersea volcanoes and thermal vents and reducing metals, minerals and carbon dioxide.

For the next billion years photosynthesis (and thus oxygen production) was not significantly hampered by any sun-blocking organic haze or the feeble rays of the sun, due to the activity of viruses (Joseph 2009b,c,d). Viruses provide bacteria with additional photosynthesizing genes under conditions of reduced sunlight (Lindell et al., 2004; Sullivan et al., 2005, 2006; Williamson et al., 2008).

Microfossils discovered in the Murchison meteorite which resemble cyanobacteria

Stromatolites

Thus, beginning nearly 4 billion years ago, oxygen-producing Cyanobacteria were proliferating and began creating thick calcium encrusted Cyanobacterial mats (Buick 1992), as well as secreting oxygen, and leaving their fossilized signatures in shales and stromatolites (Brocks et al., 1999; Olson 2006). And by 3.8 bya they were leaving organic carbon and grains of calcium in their wake-- the residue of photosynthesis, oxygen secretion, and thus biological activity as discovered in quartz-pyroxene rocks on Akilia, West Greenland, (Manning et al. 2006; Mojzsis et al. 1996). Around 3.2 bya, there was a spike in atmospheric oxygen, a consequence of increased oxygen photosynthesis (Ohmoto et al. 2005). By 2.45 bya, oxygenic photosynthesis had become widespread (Brock et al., 2003; Buick 2008) and atmospheric oxygen levels rose (Bau et al. 1999; Kirschvink et al. 2000) to values between 0.02 and 0.04atm (Holland 2006). In consequence numerous microbial species perished, but in dying became food sources for species which had just begun to evolve.

Makganyene Glaciation, Metamorphosis, First Snow Ball Earth

The Proterozoic Eon (2.5 bya – 542 mya), was punctuated by four major cycles of global cooling and world wide glaciation, which nearly froze the planet and creating three episodes of what has been called "Snow Ball Earth." Much of this global cooling was a consequence of biological activity; i.e. oxygen production by photosynthesizing Cyanobacteria, and methane consumption by methanotrophs (Joseph 2009d,e). However, it should be stressed that solar and geomagnetic activity, coupled with variations in the orbit and tilt of this planet, were also contributory factors.

Because of the high levels of methane which had built up in between 2.8 to 2.5 bya, archae known as methanotrophs and methylotrophs began to proliferate. These were methane eaters, and in ever growing numbers they metabolized and broke down methane, as demonstrated by the presence of hopanes and high relative concentrations of 2α-methylhopanes in Archean rocks (Brocks et al., 2003). As methanotrophs proliferated, methane levels rapidly fell, thereby reducing the green house effect which also allowed more sunlight to strike Earth. Increased sunlight triggered increased photosynthesis. By 2.45 bya, oxygenic photosynthesis had become widespread (Brock et al., 2003; Buick 2008) and atmospheric oxygen levels rose (Bau et al. 1999; Kirschvink et al. 2000) to values between 0.02 and 0.04atm (Holland 2006).

Oxygen also breaks down methane. Indeed, the presence of even small amounts of O_2 in the atmosphere would have been associated with a significant decrease in its CH_4 content, and this decrease would have caused the planet to rapidly cool (Young et al. 1998; Kasting & Ono 2006). In fact, O_2 levels became so high around 2.4 bya that sulphur MIF production collapsed, and this caused a rapid and drastic decrease in atmospheric CH_4, thus triggering glaciation (Kasting and Howard, 2006). That is, increased levels of O_2 acted to oxidize sulphide, such that dissolved sulphate levels increased just as O_2 levels increased. Both began to build up in shallow marine sediments which resulted in decreases in methagenesis and significant reductions in atmospheric methane (Pavlov et al. 2003; Kharecha et al. 2005). The increased levels of sulphate in turn triggered a proliferation of sulfur-eating bacteria, which caused a drawdown in H_2 and CH_4, a consequence of bacterial sulphate reduction (Kasting and Ono, 2006).

Because the sun's output was only about 90% of current levels, the loss of this heat trapping greenhouse methane and carbon dioxide blanket caused the planet to rapidly cool.

Moreover, CO_2 levels were being reduced by photosynthetic bacteria who were employing H_2, H_2S and/or Fe^2+M to reduce CO_2 to organic matter (Pierson 1994). The reductions in methane coupled with reductions in CO_2 accelerated the cooling and glaciation of the planet.

Thus, between 2.4 bya to 2.2 bya, as oxygen levels rose, the greenhouse effect was eliminated, and the planet grew cold and began to freeze (Joseph 2009d,e; Roscoe 1969, 1973), creating the first "Snowball Earth" referred to as the "Makganyene" glaciation. By 2.2 bya much of Earth and its oceans were frozen or covered with ice and snow (Evans et al., 1997; Kirschvink, et al. 2000; Roscoe 1969, 1973).

The first Snow Ball Earth was orchestrated biologically. The changing environment then acted on gene selection, triggering the next stage of metamorphosis as well as ushering in waves of mass extinction; "pruning" of the tree of life.

Throughout the "Makganyene" glaciation blankets of snow and layers of ice also provided protection against UV rays for those living beneath the surface, but allowed light penetration (McKay 2000). This enabled photosynthesizing Cyanobacteria to proliferate (Cockell et al. 2002; Cockell & Cordoba-Jabonero 2004). These subsurface photosynthesizers secreted even more oxygen into the atmosphere, thus maintaining these freezing temperatures.

And then temperatures began to rise and the planet warmed. Yet another wave of mass extinction coupled with the evolution of new species ensued. However, those who died again became sources of food, thus providing for those who were yet to evolve.

Metamorphosis

Due to biologically engineered changes in the gas composition of the atmosphere, the planet began to warm and the first global ice age came to an end. Climate change, oxygenation, oxidation, and numerous other factors all acted on gene selection, such that, beginning around 1.8 to 1.6 bya there was an exponential explosion of diverse DNA-based eukaryotic life across the planet and within its seas (Dyall & Johnson 2000; Hedges et al. 2001, 2004; Hedges & Kumar, 1999; Wang et al. 1999). Some Eukaryotes soon consisted of approximately 10 different cell types (Hedges et al. 2004) and included organic-walled acritarchs who had not yet evolved any form of ornamentation (Li et al. 1995, 1998; Wan et al. 2003; Yan & Liu 1993).

This increase in size and complexity was made possible by the abundant carbohydrate-rich food sources available, consisting of layers of dead microbes, as well as the energy provided by mitochondria which used oxygen as an energy source. The ample supply of biologically produced nitrates and carbohydrates which could be converted to amino acids were utilized to expand the genome. Thus, those who died provided organic residue, nitrates, carbohydrates and other substances which provided the energy and the elements necessary for duplicating individual genes and entire genomes.

As genes act on the environment which acts on gene selection, additional genes were activated, genomes were duplicated, genes were discarded, and new functions, characteristics, and species began to appear whereas others became extinct. However, not just the Eukaryotic genome was impacted, but the mitochondria genome (Rogers et al. 2007). Mitochondria and a variety of Prokaryotes including Cyanobacteria donated numerous genes which were integrated into the Eukaryotic genome. These included genes coding for organelles and the endoplasmic reticulum, as well as genes contributing to the nucleus, and the bacterial-type plasma membrane that displaced the original archaeal membrane (Esser et al. 2004; Rivera & Lake 2004); a process Andersson (2005) refers to as "endosymbiotic gene transfer."

The endoplasmic reticulum, which is related to Cyanobacteria HGT, also processes the calcium secreted by Cyanobacteria, and then synthesizes collagen which stimulates calcium binding, thereby setting the stage for the metamorphosis of bones, the skeletal system, and the brain one billion years later, around 540 million years ago.

Some Cyanobacteria also invaded these ancestral plant-Eukaryotes and then underwent a genetic metamorphosis and were transformed into organelles, the chloroplasts (responsible for photosynthesis, fatty acid synthesis, and the plant's immune response). Chloroplasts and Cyanobacteria clearly resemble one another functionally and physically, and share identical genes (Joyard et al., 1991; Martin et al., 2002; Krause 2008; Keeling 2004).

By 1.6 bya the genome of photosynthesizing Eukaryotes again duplicated in size (Alvarez-Buylla et al. 2000), thanks in part to waves of viral invaders and the insertion of viral regulatory genes (Joseph 2009c,d). In consequence, plants and animals diverged from all possible common ancestors (Wang et al. 1999). Later, many species of plant would become food sources for animals; and in this regard, this divergence is reminiscent of the splitting of the pre-embryo, part of which becomes a food source which nourishes the part which becomes the embryo.

The genes and genomes of plants and animals underwent repeated duplications. In plants, these whole genome duplicative events created multiple copies of MADS-box genes (Alvarez-Buylla et al. 2000) which over a billion years later would regulate the expression of flower, fruit, leaf, and root development (Ng and Yanofsky 2001; Pelaz et al. 2000). Whole genome duplication in the plant lineage would be followed by a number of recombination genetic events creating new plant-gene sequences from old genes coupled with gene loss and deletion (Alvarez-Buylla et al. 2000).

Around 130 million years ago, leafless plants would begin to flower and fruit, thereby providing additional food sources for numerous species of animal, some of which dined on roots, others on leaves, and yet others on the fruit.

Calcium, Cerebrums, & Metazoan Multicellularity

Calcium plays a central role in cellular proliferation and differentiation, cell to cell adhesion and fusion, and apoptosis, programmed cell death (Brown and MacLeod 2001; Cheng et al. 2007). In the absence of Ca, cells stop aggregating, embryos fail to adhere, cell aggregates disintegrate, and bones become soft and easily break.

Following the end of the Marinoan/Gaskiers glaciation, and as temperatures rose, calcium flooded the oceans due to coral reef and stromatolite evaporation. The rapid and massive increase in calcium levels acted on gene selection and triggered a whole spectrum of calcium binding, and calcium-collagen proteins activities, including the creation of the skeletal system (Joseph 2009a).

Cells absorb and secrete Ca^2+, and calcium receptors and sensors are located throughout the body and the muscular-skeletal system, including on cartilage and bone cells (Brown & MacLeod 2001; Chang et al. 1999; Cheng et al. 2007). Calcium binding proteins also regulate many important cellular processes such as smooth muscle contraction and motion in skeletal muscle.

Hence, calcium plays a key role in the skeletal muscular system, and the regulation of cell, muscle, and skeletal functioning in metazoans. The buildup of calcium was central to the metamorphosis of macro-multicellular Eukaryotes which diversified and increased in size following the end of the Marinoan glaciation followed by the Gaskiers glaciation around 580 mya.

During the warming period following the Marinoan glaciation (650 to 635 mya), Cyanobacteria colonial mats which were glued together with calcium carbonate secreted by Cyanobacteria, began to decompose and calcium was released into the ocean creating "calcite seas" (Hardie 2003). Thus, as early as 635 mya, a number of taxa were already displaying calcium carbonate mineralization. These included sponges who began to evolve a silica-collagen skeletal system, which included calcium, thereby forming soft, lacelike silica skeletons, spicules, and spines which enabled them to enlarge their cell walls and grow in size (Gehling and Rigby 1996; Li, et al. 1998; Tiwari et al. 2000; Xiao et al. 2000). And as they became larger, they were invaded by Cyanobacteria which had created the conditons which promoted their growth.

Following the end of the Gaskiers glaciation (which endured from 585 to 580 mya), calcium-enriched mats and reefs created by Cyanobacteria and corals (with their Cyanobacterial genes) continued to evaporate flooding the oceans with additional levels of calcium, and a complex variety of bilateral forms began to appear (Bowring et al. 2003; Grotzinger et al. 1995; Martin et al. 2000).

By 580 mya, oxygen levels had also risen sufficiently so as to generate a thin ozone layer. As ozone was being established it began to block out life-neutralizing UV rays and multi-cellular Eukaryotes became larger and composed of additional cell types. In the absence of ozone, larger sized bodies would be burnt by UV rays and would pop and explode. Thus, until around 580 million years ago, the vast majority of life forms sojourning on Earth and beneath the seas, were single celled organisms and simple multi-celled creatures composed of less than 11 different cell types (Bottjer et al., 2006; Glaessner, et al. 1988; Narbonne 2005; Narbonne & Gehling 2003; Shen et al., 2008).

Calcium and Ca^2+ ions promote skeletal and brain development, and interact with genes which code for functions mediated by the central nervous system (Glezer et al. 1999; Hong et al. 2000; Llinás et al. 2007; Köhler et al. 1996; Mori et al. 1991; Perez-Reyes 2003; Weisenhorn 1999). Hence, following the end of the Marinoan and then the Gaskiers glaciation and the flooding of the oceans with calcium, evidence of horizontal and vertical burrowing appears around 545 mya (Erwin and Davidson 2002) and metazoans began displaying evidence of behavior guided by a brain.

However, increased intracellular Ca can also trigger apoptosis (Mattson and Chan 2003) and thus cell death. Increased calcium levels possibly contributed to the eradication of entire species (Joseph 2009e). Thus, as metazoans began to proliferate, yet other species suffered a progressive and massive die off, including the Ediacaran fauna which became extinct, bringing the Ediacaran era to a close. The dead Ediacaran, in turn, became sources of food for those who were still evolving.

Once calcium levels and other substances built up sufficiently the changing environments acted on gene selection and triggered explosive bursts of evolutionary developments, giving rise to animals sporting shells, exoskeletons, bilateral bodies, bones and complex brains--a function of the massive amounts of oxygen, carbon, calcium, zinc, copper, and other liberated minerals and gasses acting on gene selection. By the onset of the Cambrian Explosion, 540 mya, an incredible variety of complex creatures began to appear.

By 540 mya, not just calcium, but so much oxygen had been released into the atmosphere that a thick ozone layer had been established. And then, all manner of complex life quite suddenly exploded upon the world stage. With the establishment of ozone innumerable creatures could emerge from the sea or from beneath the soil and exploit new environments; environments which acted on gene selection giving rise to new capabilities and new species.

Thus, it took billions of years for oxygen levels to build up sufficiently so as trigger the development of the radiation blocking ozone layer and this was made possible by Cyanobacteria biological activity. The same is true of calcium which was also secreted by colonies of Cyanobacteria to glue together their colonial mats. It was only when sufficient oxygen, ozone, and calcium had built up in the atmosphere and the seas that multi-cellular life could expand in size and emerge from the oceans, equipped with bones made of calcium and skeletons to protect brains and internal organs.

Therefore, not only had Cyanobacteria contributed many of these genes to the Eukaryotic genome, but these microbes provided the calcium and oxygen and other substances which would act on those genes thereby giving rise to bony brainy animals who could go forth and conquer the world.

Biological Environmental Engineering: Silica, Calcium, Cyanobacteria

Microbes make up the majority of the living biomass on Earth, and from the very beginning bacteria and archae have labored to genetically engineer the environment in preparation for the next stage of metamorphosis. And they did this by converting minerals, enzymes, gasses, nitrogen, sulfur, iron, hydrogen and sunlight, into forms which could be used to sustain the life of more complex creatures which had not yet evolved (Joseph 2000a, 2009b,c,d).

The creation of a changing biosphere, what could be likened to the "womb of the planet" was not hap hazard, but under strict genetic regulatory control. Most of the minerals, elements, and gasses were biochemically liberated or oxidized in an orderly, logical sequential, seemingly step-wise progression over a billion years of time, which, in conjunction with geochemical events, impacted and paralleled the evolution of increasingly complex and intelligent species (Hazen et al., Joseph 2000a, 2008, 2009d; Williams 2007).

Beginning around 850 to 820 MA, the pre-Pangean supercontinent named "Rodinia", which occupied the tropical equatorial regions, began to slowly break apart; a consequence of the biological digestion of rock and earth by microbes (Joseph 2009d), as well as plate tectonics, mantle subduction, extensive volcanism coupled with magma super plumes (Druschke et al., 2006; Li et al. 2003; Sung et al., 2006; Wang and Li 2002; Zhou et al., 2002).

Over the next 100 million years, as Rodinia continued to fracture and drift apart, greater masses of formerly very dry land were increasingly exposed to greater amounts of moisture and ocean water (Johnson et al., 2005). Microbial activity also increased and the chemical composition of the soil continued to undergo severe and rapid weathering and biological deterioration. The combined effects of microbe and weathering resulted in the release of a variety of minerals and other elements, including Zn^{2+}, Mn^{2+}, Fe^{2+} and a variety of carbonate aerosols, as well as massive amounts of silicates that had been liberated from the soil (Joseph 2009d,e). The silicates bled into rivers and drained into the seas.

Cyanobacteria secrete calcium to glue together their colonies, forming great mats and stromatolites. However, with each cycle of global cooling followed by global warming, these great mats, which had been forming for billions of years, began to decompose releasing vast amount of calcium, and calcium carbonate into the oceans which in turn activated Ca^{2+} binding proteins and genes.

There are numerous proteins that respond to silica, Zn^{2+}, Mn^{2+}, Fe^{2+} and Ca^{2+}, and which form a homeostatic link such that they bind together thereby inducing gene expression (Dupont et al. 2006; Morgan et al. 2004; Williams & Fraústo da Silva 2006; Williams 2007). As silica, Ca^{2+}, Zn^{2+}, Mn^{2+}, Fe^{2+} levels increased, they acted on specific proteins and genes, triggering their expression and possibly inducing several whole genome duplication events which increased the number of genes that could produce a greater number of Ca^{2+}, Zn^{2+}, Mn^{2+}, Fe^{2+} proteins.

Silica, which microbes had labored to mine from rocks and earth, interacts with calcium carbonate, and together the carbonate–silicate cycle directly impacts climate and ocean water chemistry (Berner et al., 1983; Berner 2004; Walker et al. 1981). Silica acts on gene selection and the massive amount of silica released resulted in siliceous biomineralization and the evolution of the Silicarea sponge, creatures which first appeared around 630 to 650 mya (Love et al., 2006; Tiwari et al., 2000; Xiao et al., 2000). The Silicarea sponge has a soft honey-comb skeleton comprised primarily of silica and collagen; collagen being directly related to calacium.

Because of increasing levels of calcium, silica, and Zn^{2+}, Mn^{2+}, Fe^{2+}, vast networks of silent genes were activated, and others silenced and an explosion of life ensued (Condon et al., 2005; Joseph, 2009d; Peterson & Butterfield 2005; Williams 2007). A new wave of speciation was unleashed, including the evolution of megascopic Ediacarans (Narbonne 2005; Narbonne and Gehling 2003). Eukaryotic life had made a giant leap from microscopic to megascopic.

Because many species of Eukaryote including the Ediacarans had grown in size (due to mitochondria, the endoplasmic reticulum, a silica skeleton and an increasingly thick ozone layer), Cyanobacteria were able to invade and form a symbiotic relationship with these species; and in so doing contributing additional genes to the Eukaryotic genome in the process, including genes related to photosynthetic activity (Cavalier-Smith 1993; Seilacher et al. 2003).

Not just Ediacarans but giant protozoa were likely invaded by Cyanobacteria, and also began engaging in photosynthesis (Cavalier-Smith 1993). As calcium carbonate is also produced as a byproduct of photosynthesis, calcium levels began to rapidly increase.

The changing environment, which included liberated silica and increasing quantities of calcium, acted on gene selection. Silica and calcium were increasingly internalized to build skeletal structures.

Silicate increases gene expression of silicatein and collagen (Krasko et al., 2001; Müller et al., 2003) creating silica spikes and a silica skeleton consisting of silica and collagen. Collagen is synthesized, in part by the endoplasmic reticulum which owes much of its origins to Cyanobacteria. However, bones, teeth, and the skeletal system require large quantities of calcium which is bound with collagen, creating a collagen-calcium-protein matrix which in turn forms bones.

Calcium, Colagen and Gene Expression

The biosynthetic pathways responsible for collagen production is exceedingly complex and are encoded and expressed by a variety of genes found on a number chromosomes. As the collagen molecule is synthesized, it undergoes many post-translational modifications which take place in the Golgi compartment of the endoplasmic reticulum, and is dependent upon peptides, calcium, and vitamin C and iron as cofactors. The endoplasmic reticular also processes calcium which promote protein folding and binding (Michalak et al., 2002) and collagen binding.

As summarized by Michalak and colleagues (2002), the endoplasmic reticulum is a centrally located organelle which affects virtually every cellular function. Its unique luminal environment consists of Ca^{2+} binding chaperones, which are involved in protein folding, post-translational modification, Ca^{2+} storage and release, and lipid synthesis and metabolism. Moreover, calcium increases the synthesis of collagen (Chen et al., 1992).

The endoplasmic reticulum owes its ancestral origins to Cyanobacteria and Cyanobacteria genes which are also responsible for generating the calcium which is then processed by the endoplasmic reticulum which also synthesizes collagen .

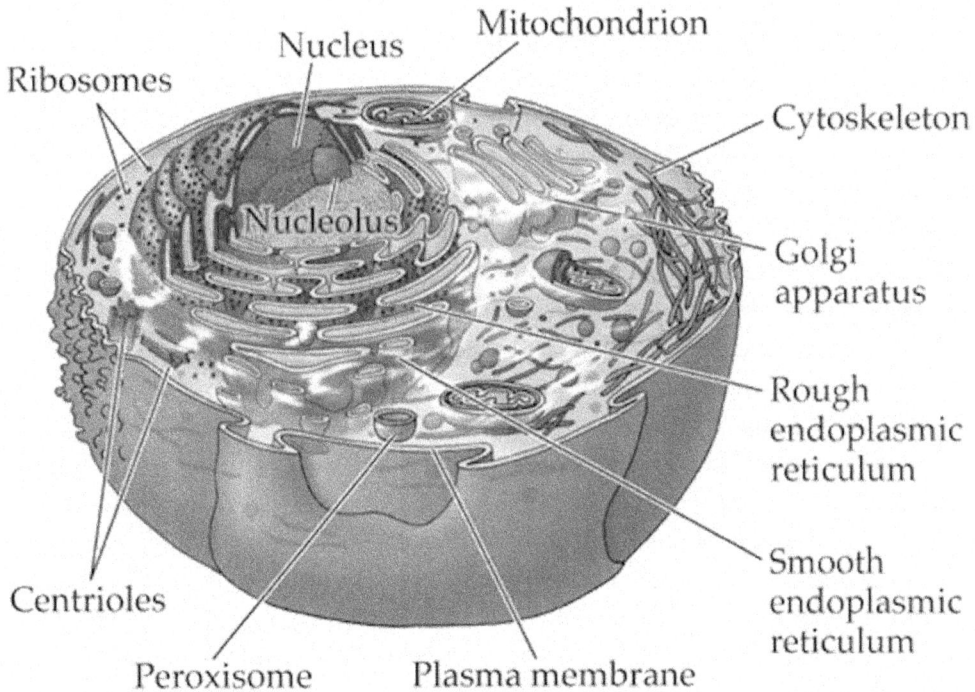

Collagen stimulates calcium binding (Chen et al., 1992) and exhibits a high affinity for calcium ions resulting in calcification and thus the creation of bones. Collagen-calcium binding also attracts phosphoproteins, and this results in the creation of bones consisting of a collagen-calcium-phosphoprotein matrix. Moreover, as calcium levels in bone increase, additional phosphoprotein bind to the collagen-calcium matrix, thus increasing the strength and elasticity of the bone.

Calreticulin, a major Ca^{2+} binding (storage) chaperone in the endoplasmic reticulum (ER), is a key component responsible for the folding of newly synthesized proteins and glycoproteins (Michalak et al., 2002). Collagen is also synthesized in the ER. The function of calreticulin and other proteins is affected by continuous fluctuations in the concentration of Ca^{2+}. Calreticulin appears to be an upstream regulator of the Ca^{2+}-dependent pathways that control cellular differentiation and/or organ development. (Michalak et al., 2002).

Thus there is a complex interaction where calcium promotes collagen synthesis, then collagen stimulates calcium binding, and then increased concentrations of calcium stimulate additional proteins to bind with the collagen-calcium matrix (Chen et al., 1992). Phosphoproteins in fact resist binding, and cannot attach to collagen or grow in the absence of calcium (Saito et al., 1998). Thus the protein matrix becomes positively charged by virtue of the bound calcium ions, which attracts neutralizing phosphate and carbonate ions, which then allow further calcium ion binding (Ury 1971) thereby creating shells, bones, and teeth.

Until sufficient oxygen, silica, and calcium had been released, body and cell size were restricted and unable to expand or engage in strenuous physical activity. Larger bodies and internal organs require skeletal support and more efficient means of energy consumption--made possible by internalized microbes which had been transformed into mitochondria which breath oxygen. All this was made possible through biological alterations of the environment and the release of silica, oxygen and calcium; the latter being directly associated with Cyanobacteria

Beginning around 640 mya and continuing for the next 60 million years, silica, calcium, and oxygen levels began to increase substantially, and then, around 580 million years ago calcium began to flood the oceans.

Cells absorb and secrete Ca^{2+} and calcium receptors are located throughout the body and the muscular-skeletal system of simple metazoans (Brown and MacLeod 2001; Cheng et al., 2007).

Ca^{2+} ions acts on gene selection, increasing the permeabilization of the inner mitochondrial membrane (Castilho et al., 1995), facilitating photophobic responses, and significantly increasing photosynthetic activity (Colombetti et al., 2008). Ca^{2+} ions therefore, can increase energy efficiency within the cell and the amount of oxygen and calcium pumped into the environment. Increased energy can also support increases in body size and complexity made possible by a calcium-collagen skeletal system.

Calcium is the most ubiquitius metal ion in the cellular system and plays a universal role as messenger and regulator of protein activities (Kazmierczak & Kempe 2004; Williams 2007). Calcium acts directly on gene expression (Castilho et al., 1995), and the regulation of programmed cell death (apoptosis), cellular proliferation and differentation, and cell to cell adhesion and fusion (Brown and MacLeod 2001; Cheng et al., 2007). In the absence of Ca cells stop aggregating, embroys fail to adhere, cell aggregates and disintegrate, and bones become soft and easily break (Kazmierczak & Kempe 2004). Therefore, until sufficient quantities of calcium had been biologically produced and then liberated, embryos and bones were an impossibility.

Brains, Calcium, Cyanobacteria, Biological Environmental Engineering:

Cyanobacteria, which were among the first to colonize Earth, had generated calcium-mats which over billions of years became hundreds of feet thick on the ocean floor. Cyanobacteria secrete calcium carbonate within their mucous (Kazmierczak and Stal 2008). These secretions are used to glue and cement stromatolites together, and to create thick Cyanobacterial mats, allowing vast colonies of Cyanobacteria to adhere to one another (Alois 2008). The lithification of these marine Cyanobacterial mats, to create rock-like sediments, is thought to be driven by metabolically-induced increases in calcium carbonate saturation (Alois 2008). The secretion of calcium carbonate to form Cyanobacteria mats, also infiltrated carbonate rocks (Alois 2008) and accelerated the mineralogy of reef-building (Porter 2006). Vast stores of calcium began to build up in these mats, stromatolites, carbonate rocks and reefs. Thus, by 600 mya, vast ocean preserves of calcium carbonate had been established.

However, these mats began to decompose and melt around 580 million years ago with the ending of the world wide Gaskiers glacial period and the ensuing warming period. Stupendous amounts of calcium were released into the oceans. Increased levels of calcium interact with silica and gene selection.

Increased levels of calcium carbonate also potentiates photosynthesis (Colombetti et al., 2008) in Eukaryotes and Prokaryotes. Increased photosynthesis increases the production and secretion of calcium carbonate (Alois 2008; Porter 2006) by Eukaryotes and Prokaryotes. Thus, an increasingly powerful feedback mechanism is put in motion where calcium carbonate potentiates photosynthesis which results in the release of more calcium carbonate as well as more oxygen.

Over the next 40 million years the protective (oxygen-initiated) ozone layer was slowly established, creatures expanded in size, diversified, and grew silica spines, silica skeletal compartments, then silica-collagen skeletons, then collagen-calcium skeletons, armor plates (sclerites) and small shells like those of brachiopods and snail-like molluscs (Joseph 2009d).

Over a period of nearly 4 billion years, until around 540 mya, calcium concentrations increased by 100,000 times (Kempe and Degens, 1985) with the greatest increases occurring during and following the Marinoan glaciation and then the

Gaskiers glaciation, 580 mya. The rapid increase in calcium levels triggered a whole spectrum of calcium binding and calcium-collagen proteins activities including the creation of the skeletal, muscular, and nervous system. Calcium binding proteins in fact regulate smooth muscle contraction and motion in skeletal muscle (Kazmierczak & Kempe 2004), and Ca^{2+} sensors are located in cartilage and bone cells that mediate some or even all of the known effects of Ca^{2+} on these cells (Brown & MacLeod 2001; Chang et al., 1999).

Hence, calcium plays a key role in the evolution and regulation of skeletal muscle movement and contraction, and thus the regulation of cell, muscle, and skeletal functioning in metazons (Kazmierczak & Kempe 2004). Once sufficient quantities had been produced, genes were activated and complex species with muscles, bones (and brains) evolved. Hence, the buildup of calcium played a central role in the creation of macro-multicellular Eukaryotes which diversified and increased in size following the end of the Marinoan/Gaskiers glaciations.

Calcium is not only a major component of the skeletal system (Nudler et al., 2003; Urbano et al., 2002), but acts on a number of genes to build and maintain the integrity of the excitable membranes of heart, glandular, and muscle cells, as well as the brain. Calcium promotes nerve cell development and plays a central role in neural generation, the functioning of the synapse, the activation of DNA which codes for neural functional organization and expression, and thus the development and functional integrity of the brain (Glezer et al., 1999; Hong et al., 2000; Llinás et al., 2007; Köhler et al., 1996; Mori et al., 1991; Perez-Reyes 2003; Weisenhorn, D. M. (1999). In fact, the calcium biomatrix obtained from the exoskeleton of the coral P. lutea has been shown to promote the morphological development of neural tissue, including astrocytes, pyramidal and granule neurons, and tissues resembling hippocampal neurons (Peretz et al., 2007; Shany et al., 2003, 2005). Ca^{2+} ions have a special affinity for genes which code for functions mediated by the central nervous system (Glezer et al., 1999; Hong et al., 2000; Llinás et al., 2007; Köhler et al., 1996; Mori et al., 1991; Perez-Reyes 2003; Weisenhorn 1999; Ubach et al., 1998).

Because ancient species passed on the necessary genes coding for the brain and nervous system, once calcium levels and other substances built up sufficiently these genes were activated in subsequent species who evolved bones, brains, and a skeletal system and a hard cranium which could protect that brain.

Increased oxygen also provided oxygenated environments throughout the ocean which could be exploited and colonized by oxygen breathing creatures. Coupled with increased calcium and silica, vast networks of silent genes were activated, and others silenced and an explosion of life ensued and a new wave of speciation was unleashed.

Cambrian Explosion and Cyanobacteria

As the planet continued to warm, and Cyanobacteria calcium mats deteriorated the oceans became increasingly saturated with calcium, creating "calcite seas" (Porter, 2006). By the onset of the Cambrian Explosion so much oxygen and calcium had been released, that beginning around 540 mya, there was a vast explosion of bilaterial metazoan diversity and complexity that appeared multi-regionally throughout the oceans of Earth within 5 to 10 million years. Over 32 phyla rapidly evolved, many with complex skeletal systems, nervous systems, hearts, "modern" eyes, and the "modern" body plans seen in present day animals (Fortey et al., 1997; Valentine et al., 1999; Conway & Morris 2000; Budd & Jensen 2000; Peterson et al. 2005). However, these species and their descendants, also biologically modified the planet and their activities effected the climate and global temperatures. Innumerable creatures emerged from the sea and from beneath the soil and exploited new environments; environments which acted on gene selection giving rise to new capabilities and increasingly intelligent species.

Thus, it was Cyanobacteria which produced or was responsible for much of the oxygen atmosphere and the calcium which enabled oxygen breathing animals to emerge from the sea with their internal organs supported and encased within a skeletal cage. And, it was also Cyanobacteria, along with archae, other bacteria, and viruses which contributed the genes to the Eukaryotic genome which responded to the increases in oxygen and calcium thereby generating bones, brains, hearts, and eyes. In fact, the single most prerequisite for the development of vision, is the vitamin-A-related chromophores in the visual pigment, and this is also found in Cyanobacteria (Seki & Vogt 1998; von Lintig & Vogt 2004), whereas the eye is an outgrowth of the brain, the evolution of which is also directly related to increased levels of calcium (Joseph 2009d).

Cyanobacteria were among the first to arrive on Earth. Fossils of Cyanobacteria and Cyanobacteria mats, have been discovered in meteors such as the Murchison (Hoover, 2011) which is older than this solar system (Joseph 2009a).

Cyanobacteria: Biological Engineered Environments Acts on Genes

Cyanobacteria, therefore, are largely responsible for generating an oxygen atmosphere which produced the radiation blocking ozone layer which allowed animals with bones, brains, hearts, and eyes, to emerge from the sea and walk upon the surface of the Earth. And where did the genes that code for hearts, eyes, bones and brains originate? Viruses and prokaryotes including Cyanobacteria transferred these genes into the Eukaryotic genome, and then Cyanobacteria produced the calcium which triggered the activation of some of these genes as well as providing the calcium-collagen matrix of which bones are comprised.

Thus, this large repertoire of silent genes were inherited from ancestral species including Cyanobacteria which were among the first to take up residence on Earth. Cyanobacteria then labored to alter the environment which acted on those genes.

Prior to their arrival, and then again, once on Earth, Cyanobacteria and other microbes (including viruses) were continually exchanging genes, and inserting genes in the eukaryotic genome. After becoming established on this planet, these genes or the genetic material for generating these genes, was inserted into the Eukaryotic genome and were passed down for over 4 billion years, albeit in a silent, dormant stage. During this same 4 billion years, Cyanobacteria and other microbes were also biologically engineering the environment, and then around 540 million years ago, during the Cambrian Explosion, thousands of species rather suddenly evolved modern eyes, as well as a heart and central nervous system seemingly ex nihilo in the absence of intermediate forms or any evidence of Darwin's "small steps."

The implications are profound. Prokaryotes and viruses donated genes to Eukaryotes and then biologically engineered changes in the environment which acted on those genes thereby producing all manner of animals life, including humans. These were not coincidental acts of chance by microbes and viruses which were acting independently and randomly, but well organized interactions which were under precise genetic regulatory control; little different from the complex interactions of numerous cells that comprise the bodies of complex organisms including humans and whose collective cellular behavior is designed to promote the health, development and growth of the entire organism. Collectively, these Prokaryotes,viruses, and Eukaryotes form a supra-organism whose metamorphosis and development is dependent upon their well choreographed interactions which are linked by DNA and which are under precise genetic regulatory control.

Silent Genes: Eyes

These are numerous examples of identical genes coding for advanced traits appearing in diverse, unrelated ancestral species who lack these characteristics and who never developed these physical features, and who instead passed on these "silent" genes to subsequent species. It is only when the environment has been sufficiently altered, and following fluctuations in temperature, oxygen levels, and available food sources, that these "silent genes" come to be activated (e.g., Appasania 2012; de Jong & Scharloo, 1976; Dykhuizen & Hart, 1980; Gibson & Hogness, 1996; Polaczyk et al., 1998; Rutherford & Lindquist, 1998; Wade et al., 1997).

It has in fact been experimentally demonstrated "that populations contain a surprising amount of unexpressed genetic variation that is capable of affecting certain typically invariant traits" and that changes in environmental conditions "can uncover this previously silent variation" (Rutherford & Lindquist, 1998 p. 341). That is, these traits are encoded into genes which exist prior to their expression and which are inherited from ancestral species and then passed down to subsequent species at which point they are activated by environmental change or genetic regulatory signals. However, the changes in the environment are not random (albeit subject to cosmic catastrophes and other world-shaking events). For the most part they are under biological genetic control. That is, the environment is altered biologically, and the changed environment acts on gene selection (Joseph 2000a, 2009b,c,d); genes which were inserted into the Eukaryotic genome by the ancestors of the same species who biologically engineered the environment which activated those genes.

Consider, for example, the eye. The genes coding for vision and the eye in humans and other mammals, such as Pax genes ("Pax-6"), have been found in the genomes of numerous ancient species including the sea urchin and Trichoplax which have no eyes and cannot see (Sodergren et al., 2007; Callaerts et al., 1997; Hadrys et al., 2005). In fact, sea urchins, Tricholplax, and humans, share genes directly related to the limbs, brain, and the visual, auditory, olfactory, and immune system (Sodergren et al., 2007; Hadrys et al., 2005) although they diverged from common ancestors anywhere from 640 mya to 1.2 bya (Nei et al., 2001; Peterson et al., 2004; Gu 1998; Wang et al., 1999). They inherited these genes from ancestors who did not have limbs, brains, or a visual, auditory, olfactory, or immune system.

Many of these genetic commonalities can be traced back to Prokaryotes and these include the same genes coding for core cellular functions, and which are also found in plants and fungi (Koonin et al., 2004; Koonin & Wolf, 2008). Their presence in the Eukaryotic and Prokaryotic genomes indicates these genes have a history on Earth that may extend over 4 billion years in time, and then beyond to other planets. These genes were then passed down from genome to genome and from species to species albeit in silent form, only to become activated, almost simultaneously, in numerous species during a ten million year period as witnessed by the onset of the Cambrian Explosion 540 mya.

Silent genes are often activated by the changing environment (Appasania 2012), and they can be expressed following experimental manipulation. For example, silent Pax-6 eye genes which code for eye structures and which are found in body structures incapable of sight, can be experimentally activated, creating cornea, pigment cells, cone cells, photoreceptors, and thus eyes in tissues and body part where eyes should never be located such as wings, legs, and antennae (Gehring 1996; Halder el al., 1995a,b; Tomarev et al., 1997).

Silent Genes: Flowering Plants and Cyanobacteria

Much of the Eukaryotic genome consists of silent genes and retro-elements, including genes contributed by Cyanobacteria; and the same is true of non-flowering plants which possess the silent genes for producing flowers. These silent genes were passed down for almost 4 billion years, from Cyanobacteria to sea weeds, to grasses, to leafy planets, and then 130 mya these genes were expressed.

Specifically, around 130 million years ago, flowering plants evolved from leafy plants which in turn evolved from non-flowering plants which first appeared over 1.6 billion years ago (Friedman 2006; Friedman et al., 2004). However, non-flowering plants possess the genes responsible for producing petals, stamens and carpels, and thus the genes for generating the flowers of flowering plants, i.e. MADS-box genes, APETALA1, and SEP genes which are directly responsible for producing petals in flowering plants. The genes which transformed flowering from non-flowering leafless plants were inherited from ancient ancestral species which did not produce flowers (Theissen et al., 2000; Ng & Yanofsky, 2001; Pelaz et al., 2000). Many of these genes can also be traced to Cyanobacteria which inserted these genes into the Eukaryotic genome where they remained "silent" for billions of years. These silent genes only became activated following environmental gene interactions around 130 mya, thereby producing flowering plants.

These silent genes can also be experimentally activated to produce flowers in non-flowering plants. Specifically, non-flowering plants (like flowering plants) contain the SEP, MADS-box, and APETALA1 genes which are responsible for producing flowers, but in a non-activated silent form. Yanofsky and colleagues were able to activate these silent genes to produce flower petals from leaves such that the leaves of flowerless plants were converted into flowers (Mandel and Yanofsky 1995; Pelaz et al., 2000, 2001). Thus, "primitive" plants which normally never produce flowers began to flower once these silent genes were experimentally activated.

And where did these genes come from? They were inherited from ancestral species who are descended from the first living creatures to arrive on Earth, including, in particular Cyanobateria. Cyanobacteria are a direct ancestor to all plants, and along with archae, contributed numerous genes to what would become the plant genome billions of years ago (Doolittle 1999; Nosenko & Bhattacharya 2007) long before the first plant had even evolved.

Cyanobateria had already begun to proliferate upon the surface of Earth by 3.8 bya as based on evidence of calcium and high levels of organic carbon--the residue of photosynthesis, oxygen secretion, and thus biological activity (Manning et al. 2006; Mojzsis et al. 1996). Cyanobacteria is the only known species of Prokaryote capable of photosynthesis and it is colonies of Cyanobacteria which produce calcium to weld together their mats. In fact, colonies of fossilized Cyanobacteria (the ancestors of plants), and fossils of *Pedomicrobium*, a flowering bacteria, have been recovered from the Murchison meteorite (Hoover, 2011; Pflug 1984) which is older than this solar system (Joseph 2009a). Therefore, it is reasonable to conclude that these flower producing genes, which originated with Cyanobacteria, are extraterrestrial in origin.

The insertion of these Cyanobacteria genes into what would become the plant genome may have taken place in successive waves with the aid of viruses. Initially, around 4 billion years ago, viruses, archae, and bacteria (including Cyanobacteria) contributed the core genes and regulatory elements to what would become the genome of multi-cellular eukaryotes (Joseph 2009b,c,d, 2010a). Then, between 2.2 to 1.6 billion years ago, Cyanobacteria began transferring and donating over a thousand of its genes to specific multi-cellular eukaryotes--the ancestors of what would become plants. Cyanobacteria not only donated genes, but formed symbiotic relations with the common ancestors for plants (Delwiche et al., 1997; Doolittle 1999; Martin et al., 2002; Nosenko & Bhattacharya 2007).

Some Cyanobacteria after invading these ancestral plant-Eukaryotes underwent a genetic metamorphosis and were transformed into organelles, the chloroplasts (responsible for photosynthesis, fatty acid synthesis, and the plant's immune response). Chloroplasts and Cyanobacteria clearly resemble one another functionally and physically and share identical genes (Joyard et al., 1991; Martin et al., 2002; Krause 2008; Keeling 2004). This Cyanobacteria metamorphosis and, over time, the repeated insertion of Cyanobacteria genes into what would become the planet genome may have begun in earnest 2.2 bya and continued until to 1.6 bya thereby producing the first ocean dwelling photosynthesizing plants, i.e. seaweeds (Zhu & Chen 1995).

This Cyanobacteria genetic invasion was accompanied by viral genes which, together, triggered the duplication of the plant's photosynthesizing genome (Alvarez-Buylla et al., 2000) and appear to be responsible for the divergence between plants and animals between 1.5 bya to 1.2 bya (Wang et al., 1999). Thus, plants and animals share genes which were originally silent, including animal SRF-like MADS domains and MADS-box genes (Alvarez-Buylla et al., 2000).

Therefore, Cyanobacteria as well as archae and viruses, provided genes and genetic elements which would be inherited by plants and animals, and which enabled the plant's genome to duplicate in size and to create multiple copies of MADS-box and other genes (Alvarez-Buylla et al., 2000); genes which regulate flower, fruit, leaf, and root development (Ng & Yanofsky 2001; Pelaz et al., 2000).

However, these root, leaf, and fruiting flower petal producing genes were not expressed at the same time, but sequentially, over ten hundred millions of years, in response to viral invasions and environmental change, i.e. seaweeds, roots, grasses, then leafs, followed by flowers and fruit around 130 million years ago.

The Genetic Seeds of Life: Metamorphosis

The "genetic seeds of life" swarm throughout the cosmos, and some of those "seeds" fell to Earth. And they exchanged genes and commingled genomes, and extracted genes from vast viral genetic libraries, many of which were inserted into Earthly Eukaryotic genomes. Genes subject to HGT included those coding for and crucial to the development of eyes, hearts, bones, brains, and flowering plants, all of which were passed down in silent form until finally they were expressed, beginning around 540 mya for metazoans, and 130 mya for flowering plants.

There is no evidence that these or other genes somehow randomly evolved in numerous species who diverged from common ancestors over a billion years ago; the idea is preposterous. Rather, these genes were inherited from ancestral species who were heartless, eyeless, boneless, brainless, and which never produced flowers though they had the genes to do so. Many of these genes, and the environmental changes which activated these genes, can be traced to viruses, archae, and Cyanobacteria, whose own ancestors hailed from innumerable extraterrestrial environments. However, once these genes were inserted into the Eukaryotic genome here on Earth, many remained silent and suppressed for almost 4 billion years until activated by biologically induced environmental changes which were accompanied by viral invasions.

The implications, first recognized and published by Joseph (1997, 2000a, 2008, 2009b,c,d) are staggering. Microbes and viruses arrived on Earth carrying vast numbers of genes which were inserted into Eukaryotes. Then these microbes and their descendants (including Eukaryotes) altered the environment which then activated these genes, thereby leading to a leaping step-wise progression of increasing complexity and intelligence, from singled cell to human--a process which could be likened to the fertilization of an ovum by a sperm and what is best described as "evolutionary embryogenesis," and "evolutionary metamorphosis."

Darwin insisted on a strict gradualism and insisted that the evolution took place extremely slowly and was the result of "infinitesimally small inherited modifications" and "numerous, successive, slight modifications." However, as is evident from an examination of the fossil record, evolution is characterized by long periods of stasis punctuated by giant leaps without intermediate forms (Gould, 2002). There is no genetic, phenotypic, morphological or fossil evidence of gradual change from one species to another or any fossil record of transitional forms acting as an evolutionary bridge between species (Bose 2013; Gould 2002; Ingles-Prieto, et al. 2013; Rabosky 2012).

Nor are these complex cellular interactions and the emergence of increasingly complex and intelligent species a function of natural selection. The environment acts on gene selection, and new species are born into a world which has been biologically prepared for them whereas other species are shed like falling leaves in autumn.

These *genetic seeds of life* germinated a tree of life, a forest of life, which blooms and blossoms in Spring, sheds its blossoms and leaves in Autumn, goes through a period of stasis in Winter, and which is followed by rapid growth and change again in the coming Spring. The pruning of the tree of life, also known as *evolutionary apoptosis* and *extinction*, is part of the cycle of continued growth and evolutionary metamorphosis: parts are shed so the supra-organism as a whole can grow, flourish and live.

"Infinitesimally small inherited modifications" may result in variations within a species, or between family members, but it does not lead to new species as Darwin claimed. The Cambrian Explosion, itself, disproves and refutes Darwin's strict gradualism and his theory is further discredited by genetics, silent genes, horizontal gene transfer, and the fossil record.

The genomes of almost all species so far studied contain "silent" genes which contain precoded traits, and these genes may be turned on and off in response to changing environmental conditions and the influences of regulatory genes and viral invasions, thereby inducing major structural changes--much like fertilization and pregnancy and the turning on and off of genes during development and mitosis and which are responsible for the transition from an embryo to a fetus to a neonate to an infant and then an adult (Joseph, 2000a, 2009b,c,d; Zaret et al. 2013). However, rather than 9 months, it took over 4 billion years to produce a human.

The evidence, based on genetics and astrobiology, demonstrates that the extra-terrestrial "genetic seeds of life" which first took root on this planet (and which continue to fall upon this planet), contained the genes and genetic material and instructions for altering the environment and for the metamorphosis of all life; the replication of creatures which long ago lived on other planets.

4. Viruses, Genetic Libraries, Evolution, Interplanetary Horizontal Gene Transfer

Extraterrestrial Viral Genetic Contributions to Evolution

Many Prokaryotic genes, once donated to Eukaryotes, were not replaced within the Prokaryotic genome, thus insuring that Eukaryotes would evolve and become equipped with compartments, nuclei, and mitochondria, whereas Prokaryotes would only become variable (Joseph 2009b). In fact, the Prokaryotic ancestors of Eukaryotes have shown an extensive loss of genes and regulatory elements including introns (Wolfe & Koonin 2013); a consequence due in part to horizontal gene transfer and the storage of these genes within viral genomes.

Viruses serve as mobile genetic storehouses, like a lending library, as they can receive as well as donate genes. Viruses can in fact encapsulate multiple genomes (Goff et al., 2012) and can act as a bridge between the Prokaryote genome and the Eukaryotic genome and enable genes to be transferred between multiple species (Joseph 2009b). Viruses can also activate those genes and even engineer the duplication of these genes and the entire genome. Viruses have guided and have helped direct the trajectory of evolution over the ages.

For example, retroelements were transferred from reptiles to mammals via invading poxviruses (Piskurek & Okada 2007). This invasion led to the formation of many mammalian genes via the activity of viral retrotranspons which in turn created families of genes which were repeatedly duplicated by at least five independent molecular events (Campillos et al. 2006). From mammals there followed the metamorphosis of primates which were targeted by additional viral invasions and gene insertions coupled with flurries of viral retro-activity proceeded by endogenous retroviral extinctions.

Viral genes can insert themselves into a variety of locations within the host genome where they may promote gene expression or duplication; and this has been shown to be true even in the human genome. Endogenous retroviral (ERV) sequences, such as promoters, enhancers, and silencers determine when and which genes should be turned on or off and play important roles in the evolution of new species.Once inserted into a host, these viral genes can also rapidly replicate and increase in number (Doolittle & Sapienza 1980; Tsitrone et al., 1999) enabling them to grain greater control over the expanding Eukaryotic genome.

Further, just as Prokaryotes have supplied Eukaryotes with "silent genes" viruses have done likewise. Retroviruses, for example, intwine their genes with the host genome and manufacture enzymes (reverse transcriptase) to reverse transcribe the host's RNA to create a complementary viral DNA which then becomes an integral part of the host genome. Numerous copies of this viral DNA and RNA are then replicated and then transmitted via the Eukaryotic germline through daughter cells, to subsequent generations and species.

Transposable element

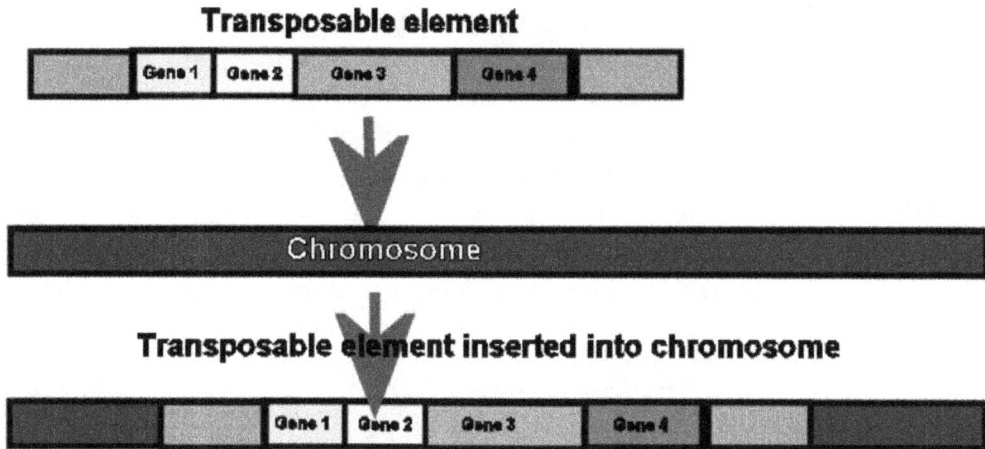

These viral genes exert regulatory control over gene and whole genome duplication, and mediate gene and genome rearrangement, transduction and the silencing vs activation of genes (Crombach & Hogeweg 2007; John & Miklos 1988); all of which can lead to new gene products and the evolution of new species. Many ERV families have lived for tens of millions of years in the mammalian genome, passed down from species to species, and then throughout the primate lineage and displaying periodic bursts of activity, such as during the split between old world and new world monkeys, the evolution of apes, and the human-chimpanzee split and leading to humans (Costas 2001; Mayer et al., 1998; Medstrand & Mager 1998; Reus et al., 2001).

For example, comparisons of gene content between macaque, human and chimpanzee genomes (Hahn et al., 2007) support an overall increase in viral and gene duplication activity in the common ancestor of chimpanzees and human. In addition, human gene duplications are genetically more diverse when compared with chimpanzee duplications (Cheng et al., 2005) which in turn is directly associated and linked to the evolution of characteristics which are distinctly human and which lend support to the prospect of continuing evolution in the human lineage (Joseph 2009c,d,e).

Eight percent of the human genome consists of around 200,000 endogenous retroviruses (IHGSC 2001; Medstrand et al., 2002), and 3 million retro elements (Medstrand et al., 2002). Some of these retroviruses are still active whereas others are silent or have been deactivated (Conley et al. 1998; Medstrand & Mager, 1998).

These, "silent" viral genes, such as those promoting the duplication of genes and whole genomes, have shown sudden bursts of activity which has corresponded with the evolution and divergence of species. Further, once a new species has been genetically manufactured, many of these viral elements become inactive or they are deleted from the genome of subsequent species. Having acted out their role in the metamorphosis of specific species, they are deleted.

For example, hundreds of deletions took place independently in the chimpanzee and human lineages after their divergence from their last common ancestor (Sen et al. 2006, Han et al. 2007). There is evidence for an almost twofold increase in gene loss in humans and chimpanzees when compared with macaques, and an almost fourfold increase in contrast to other mammals including dogs, mice and rats (Hahn et al., 2007; Rhesus Macaque Genome Sequencing and Analysis Consortium, et al., 2007; Wang et al., 2006). However, at the same time the human and ancestral hominid genome have been subject to dozens of subsequent viral invasions.

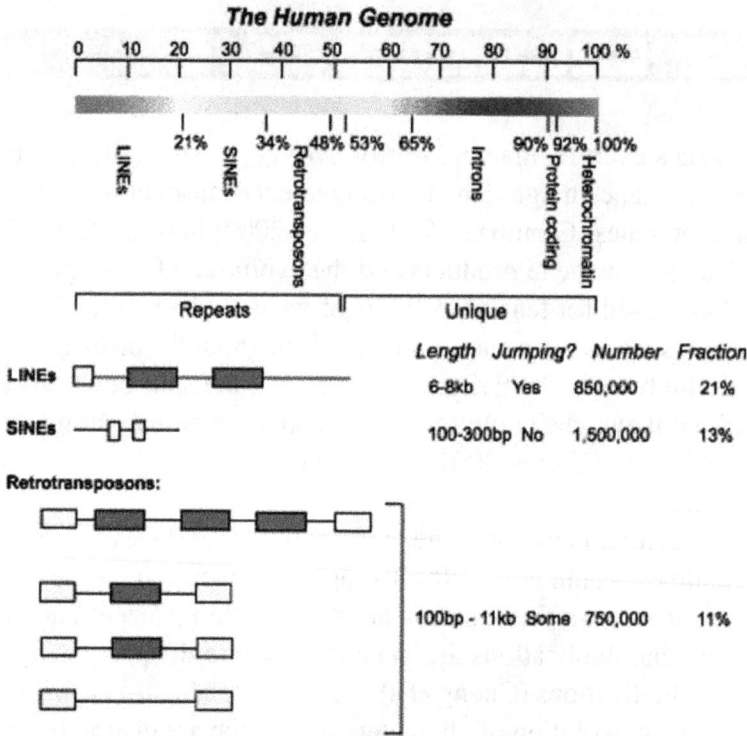

Several endogenous retroviruses (ERV) families are still active in present-day humans (Belshaw et al., 2005; Löwer et al., 1993; Medstrand & Mager 1998) which suggests that evolutionary activity will not stop with modern humans. Retroviral sequences encode tens-of-thousands of active promoters and thus regulate human transcription on a massive scale (Conley et al., 2008). In fact, about one quarter of all analyzed human promoter regions harbor sequences derived from viral elements (Jordan et al. 2003). Coupled with the 158,000 mammalian retrotransposons inherited from common ancestors, genome sequencing reveals that 8% of the "modern" human genome consists of human endogenous retroviruses (HERVs); and, if we extend this to HERV fragments and derivatives, the retroviral legacy amounts to roughly half of Homo sapien DNA (Bannert & Kurth 2005; Medstrand et al., 2002). Thus, viral genes have accumulated in those genomes leading to humans and have increased after the metamorphosis of humans; which again suggests that evolution of the human lineage may continue well in the future.

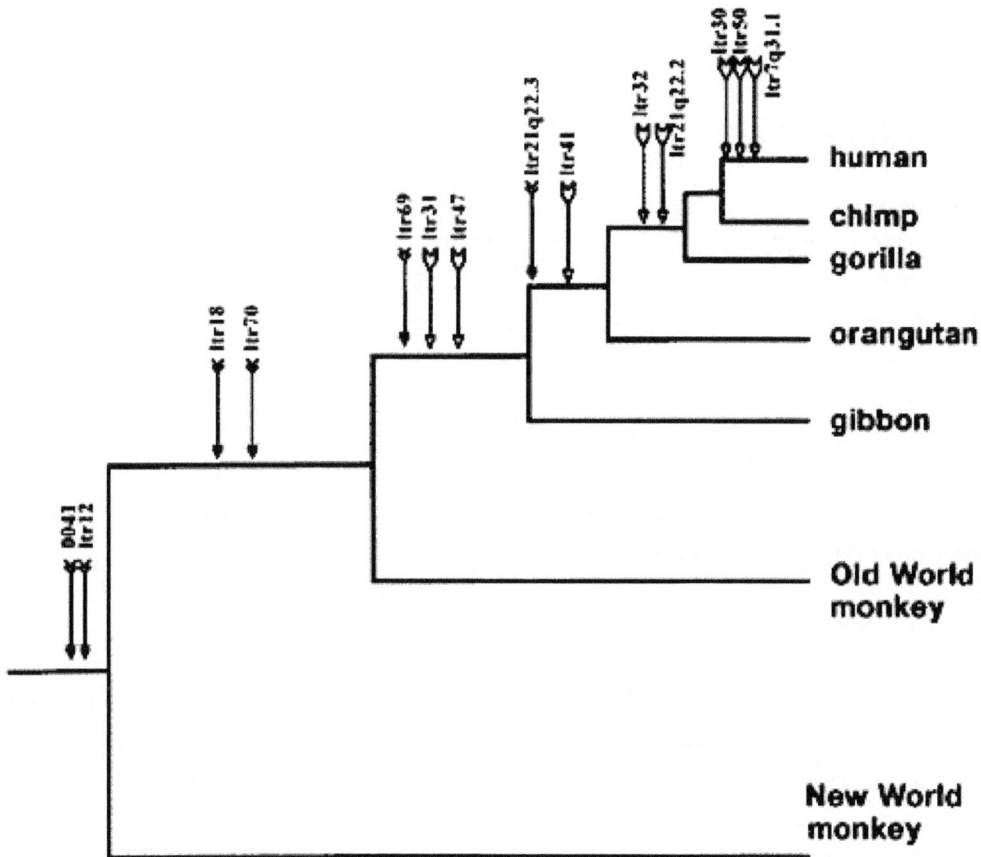

Endogenous viruses have impacted primate evolution.

Like other species, human evolution has been shaped by successive waves of viral invasion which have induced large-scale gene deletions, gene duplications and chromosome reshuffling, thereby pruning away or turning off unnecessary genes and providing additional genes and gene families. Genes have been turned on and off, genes have been split, different sequences of exons have been activated or silenced, and genes have been combined and then expressed giving rise to quantum leaps in evolutionary metamorphosis; and every stage is linked to viral activity. Viruses and have been a major source of genetic diversity (Hughes & Coffin 2004).

Thus Prokaryotes and genes contributed by Prokaryotes to the Eukaryotic genome have effected the environment which triggers the activation of these genes. Viruses and endogenous retroviruses, have profoundly affected individual genes and the entire genomes of numerous species in the evolutionary lineage leading to all manner of species including Homo sapiens (Mayer & Meese 2005). These interactions should not be viewed as coincidence or random acts of chance but highly regulated and precisely choreographed and in fact little different from the complex interactions which normally take place within every cell of the human body. If the illusion of viral Prokaryote individuality is rejected and if viewed from the perspective of examining the cellular interactions of a supra-organism then it becomes evident that this dance of life follows all the rules which govern basic biology.

Evolution and extinction are also linked and follow basic rules governing cellular biology, embryology and metamorphosis. Cells are shed, genes are deleted, others are turned on, species become extinct and the tree of life continues to grow and prosper. In fact, viruses also transfer genes and introns from species to species, thus insuring that some species evolve and others remain in stasis or even become extinct (Joseph 2009b,c,d,e).

And like those Prokaryotes who were among the first to arrive on Earth, these viruses or their viral ancestors, obtained many of their genes through extraterrestrial and interplanetary horizontal gene transfer, including from species which had long ago evolved on other worlds.

Interplanetary Horizontal Gene Transfer

When archae, bacteria, and (possibly) single celled Eukaryotes arrived on Earth, they were accompanied by viruses with their vast stores of genes which had been acquired via horizontal gene transfer (HGT). Extraterrestrial archae, bacteria, and viruses, contained the genes, genomes, and genetic information which not only contributed to the generation of the first multi-cellular Eukaryotes, but for altering the environment, and shaping and guiding the "evolution" and the metamorphosis of every creature which has walked, crawled, swam, or slithered upon the Earth, including humans.

Bacteria, archae, and viruses are ideally suited for acquiring and making copies of genes, transferring these genes to each other and to other species, as well as accepting foreign genes, and then later donating and transferring these genes, including their own genes, to yet other organisms (Forterre 2006; Hotopp et al., 2007; Iyer et al., 2006; Nikoh et al., 2008). Prokaryotes and viruses have transferred numerous genes to each other as well as to the Eukaryotic genome over the course of the history of this planet, thereby promoting multi-cellular evolution and the genetic-biological engineering of the planet.

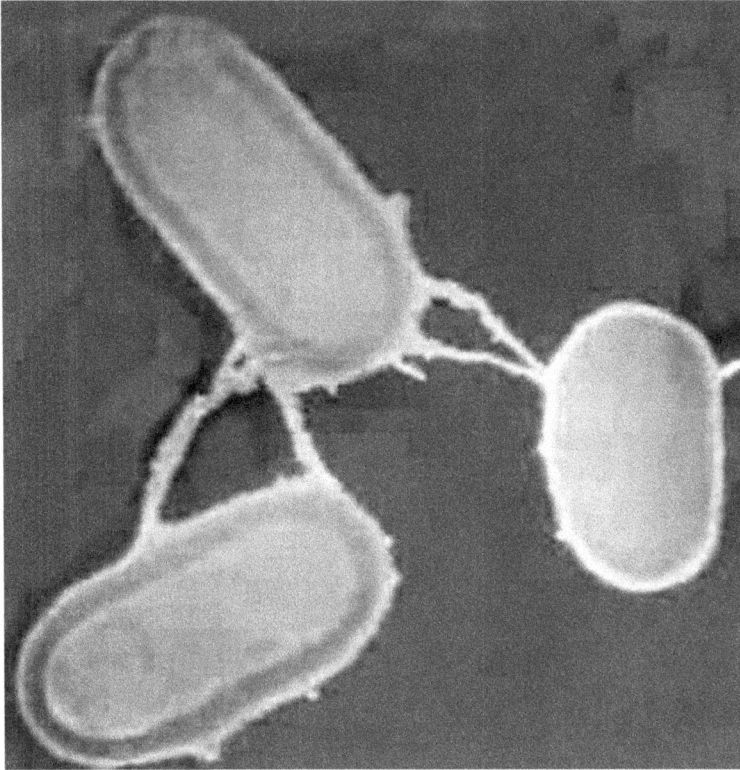

Given that it would take at least 10 billion years to generate a genome sufficient to maintain the life of the simplest organism, beginning with a single gene,
coupled with the fact that essential ingredients for the creation of proteins, DNA,
and life were not available on the young Earth, and that radiation and thermal motion would have destroyed all biological and pre-biological molecules, it can be
concluded that the ancestors of these viruses, Prokaryotes and their genes came
from other planets and journeyed throughout the cosmos encased in meteors, asteroids, comets, and planetary debris (Joseph 1997, 2000a, 2008, 2009a). And
each time these microbes came into contact or were hurled upon another planet,
they exchanged genes with each other and with any life forms already dwelling
on those planets. Life's genetic ancestry leads interminably into the long ago and
to innumerable planets and extraterrestrial environments.

Using Earth as an illustrative example, it can be deduced that horizontal gene
exchange is a cosmic imperative and has taken place on every planet and every
extraterrestrial environment capable of sustaining life. Therefore, when Prokaryotes and viruses arrived on Earth, they carried within them genes and genomes
which had been acquired or copied from creatures which had evolved on other
more ancient worlds.

If we accept the creationist theory of a "Big Bang", which was first proposed by the Catholic priest Lemaître (1927, 1931a,b) to make the Bible scientific, and current estimates as to the age of the universe, then life has had at least 13 billion years to become established throughout the cosmos and at least 9 billion years to evolve on other worlds before Earth became a member of this solar system.

If instead the universe is eternal and infinite, then life had infinite time to be distributed on every planet and moon in existence, and infinite time to evolve.

Be they formed in the same or completely different extraterrestrial environments, once archae, bacteria, and viruses came into contact, they would have exchanged genes via HGT. When their descendants were cast upon other worlds via mechanisms of panspermia (Joseph & Schild 2010b), HGT would have continued. Over the eons, as these genetic seeds of life swarmed throughout the cosmos and fell upon additional planets, HGT would take place continually between these arrivals from space and various species already evolving on these other worlds. HGT would take place continually on other planets, augmented by the arrival of additional microbes and viruses from space.

And when prokaryotes and viruses were ejected from the surface of other planets and into space following bolide impact or due to powerful solar winds, the descendants of the survivors would eventually land on other worlds carrying in their genomes numerous genes obtained via HGT and from advanced life forms whose own genetic ancestry can be traced through HGT and leads into the long ago.

It can be predicted that interplanetary horizontal gene transfer is a cosmic imperative of life throughout the cosmos. HGT is not limited to life on Earth.

Where would all these accumulating genes be stored? Within a nearly infinite variety of viruses (Joseph 2009b,c,d,e), some of which can store multiple genomes (Goff et al., 2012).

Initially, after life was first formed and spread throughout the cosmos and eventually landing on habitable planets, horizontal gene exchange would have taken place continually between archae, bacteria, and viruses. If these extraterrestrial Prokaryotes and viruses were not already accompanied by single celled Eukaryotes, then it can be predicted that viruses and Prokaryotes commingled genes thereby fashioning a genome which led to single celled then multi-cellular eukaryotes. Therefore, using Earth as an example, it can be predicted that on every habitable Earth-like planet, Eukaryotes would have evolved.

But initially, there would have been innumerable failed and successful experiments in species evolution (Joseph 2000a). Be it 10 billion years ago, or infinitely long ago, extraterrestrial experiments and failures in evolution would have occurred on countless planets; influenced and shaped by the transfer of genes inserted by viruses, archae, and bacteria which arrived from space, and by the unique and changing environments of these worlds.

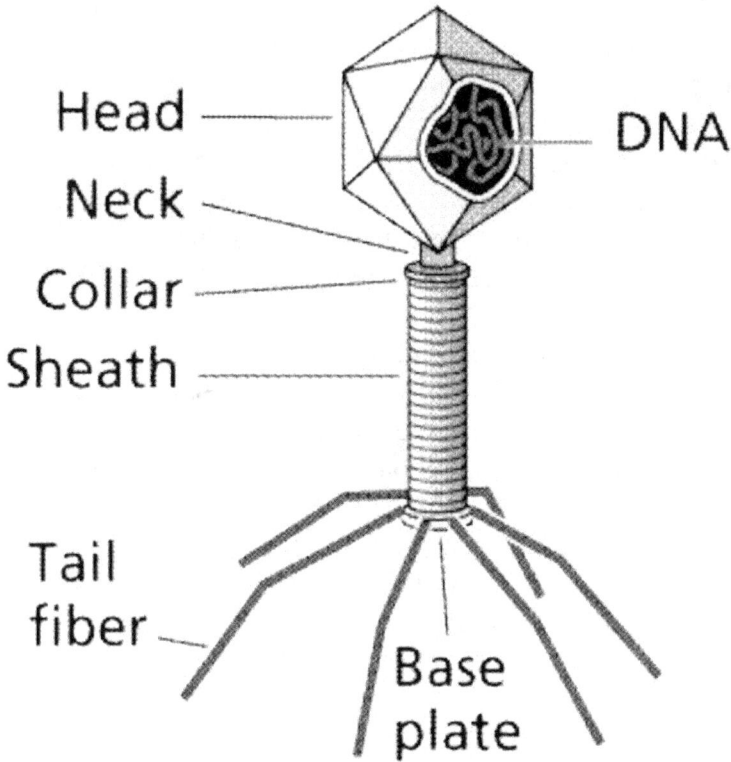

Virus

When they fell upon new worlds, Prokaryotes, Cyanobacteria in particular, would have begun digesting planetary resources, as well as each other (and perhaps any suitable life forms already extant on these alien words), and in so doing, releasing methane, oxygen, calcium, and other metabolic products (Joseph 2009b,c,d). The build up of these secretions, excretions, and other products, such as oxygen and calcium, would have significantly altered the environments of habitable planets thereby activating and expressing silent genes which code for advanced traits. Eventually, specific genes and gene combination would have enabled various species to breath oxygen, and develop hearts, bones and brains. The genes and genetic information corresponding to and responsible for these evolutionary and environmental changes would also be stored in these viral genetic libraries (Joseph 2009b,c,d)

If we ignore the evidence favoring an infinite universe, and pick a hypothetical birth date of 13.6 bya for the beginning of life, and using the evolution of life on Earth as an example, then it could also be predicted that sentient, intelligent life would have evolved on numerous Earth-life planets by 9 bya (Joseph 2000a, 2011; Joseph & Schild 2010a). This could mean that the genetic template for the evolution of all manner of life, including those similar to humans, would have been established almost 5 billion years before Earth became Earth.

And as archae, bacteria, and their viral luggage journeyed from planet to planet and solar system to solar system, they would have carried with them these genetic templates and the genes and genetic instructions for recreating these experiments in evolution; genes which would enable them to adapt to almost any environment, and if possible, to biologically and genetically engineer those environments which would then act on gene selection, such that the genetic templates coding for various life forms would come to be expressed. And this is how life on Earth originated, and why it has evolved (Joseph 1996, 2000a, 2008, 2009a,b,c,d).

Viruses, HGT, and the Evolution of Life From Other Planets

It can be predicted that initially, be it 13 billion years ago, or continually throughout infinite time infinitely long ago, that a variety of genetic experiments in evolution took place, on millions, billions, and trillions of Earth-like planets, long before the birth of our solar system. As variations in nucleotide sequences, coupled with individual and whole genome duplication as well as HGT and gene deletions, gene reorganization, etc, gave rise to new gene combinations and varying gene products and species, then the viral genetic libraries storing these genes, genomes and genetic instructions for all these evolutionary experiments would continually expand.

Long before life took root on Earth, viruses as well as Prokaryotes would have played a major role in the evolution of life on other planets, and to each successful and each failed evolutionary experiment, including the creation of new gene combinations and sequences of nucleotides, and the regulation of gene expression giving rise to a multitude of species. Billions of years later, descendants of these viruses and Prokaryotes fell to Earth and the "genetic seeds" within their genomes took root, and then they began to evolve.

Viral Contributions to Evolutionary Metamorphosis

It is well established that viruses have inserted regulatory elements which control gene expression and individual gene and whole genome duplication in Eukaryotes, thereby enabling the gene pool to grow in size; all of which leads to evolutionary innovation and speciation. Viral genes once inserted in various locations within the host genome can promote and enhance gene expression or the silencing of genes, and can help determine when and which genes should be turned on or off. Viral genes also promote various gene combinations which in turn can produce any number of products and species.

For example, it has been well established that viruses have directly inserted regulatory elements into the Prokaryotic and Eukaryotic genome, including the genome of humans, such as group II introns and ribozymes (Dai & Zimmerly 2003; Dai et al., 2003).

Introns are of particular importance in regulating gene expression (Dieci et al., 2009; Lai, et al., 1998; Noe et al., 2003; Shabalina et al., 2010) and play a major role in the regulation of gene transcription and the creation of new genes from old genes. Introns,within segments of exons, determine which nucleotide segments are expressed. If different "starter" or "stop" introns are activated this results in different segments or nucleotide sequence lengths becoming expressed, thereby producing a different protein product (Belfort, 1991, 1993; Dieci et al., 2009; Leff et al., 1986; Shabalina et al., 2010) which can give rise to different tissues and organs. Hence, variation, diversity, and perhaps even the evolution of other species can be differentially induced if different "stop" "starter" or promoter introns are activated or deactivated.

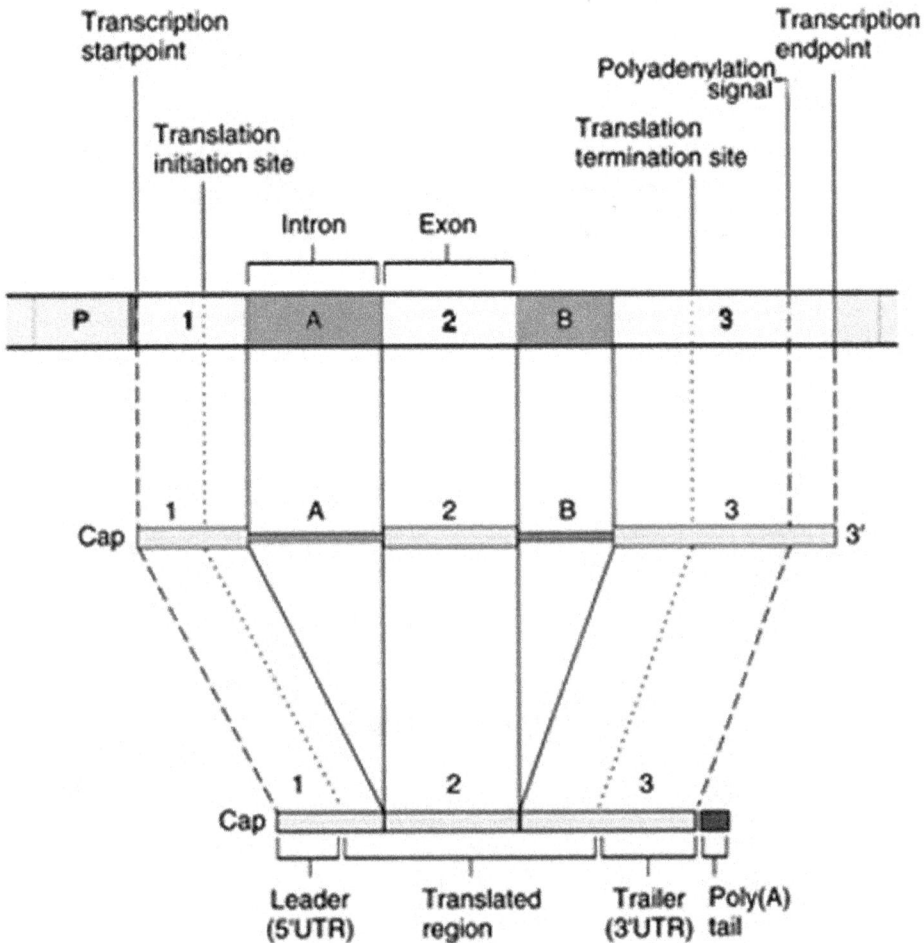

In fact, retroviruses have continually inserted genes and introns into Eukaryotes over the history of this planet, and these invasions have continued with the primate lineage leading to humans (IHGSC 2001; López-Sánchez et al., 2005; Medstrand et al., 2002; Romano et al., 2007). Coupled with other evidence, this suggests that viruses may have served as a genetic storehouse for introns which were subsequently and sequentially, over time, transferred to Eukaryotes when these hosts evolved, thereby directly influencing the trajectory of evolution leading to humans and other species.

Viruses can also insert introns as well as genes into bacteria and archae. In bacteria, approximately 35% of group II introns are linked to viruses and plasmids and are thus highly mobile (Dai et al., 2003; Klein & Dunny 2002). Introns can exit the bacteria genome and insert themselves into the genomes of Eukaryotes, other Prokaryotes, and possibly back into the viral genome. Plasmids and viruses share many characteristics as both serve as mobile carriers of packets of DNA.

Viruses are a major source of introns. Group II introns are highly mobile retro-elements (Belfort et al., 2002. Lambowitz et al., 1999; Lambowitz and Zimmerly 2004) and include retrotransposons (Beauregard, et al., 2008) and are progenitors of nuclear spliceosomal introns (Cavalier-Smith 1991; Jacquier 1990; Sharp 1991); which serve to splice gene and nucleotides together. Retroelements have as their source, retro-viruses. These retroelements can also splice together exons which are the coding portions of a sequence of nucleotides (Bonen & Vogel 2001; Michel & Ferat 1995). By splicing together exons, these viral elements can create genes from genes.

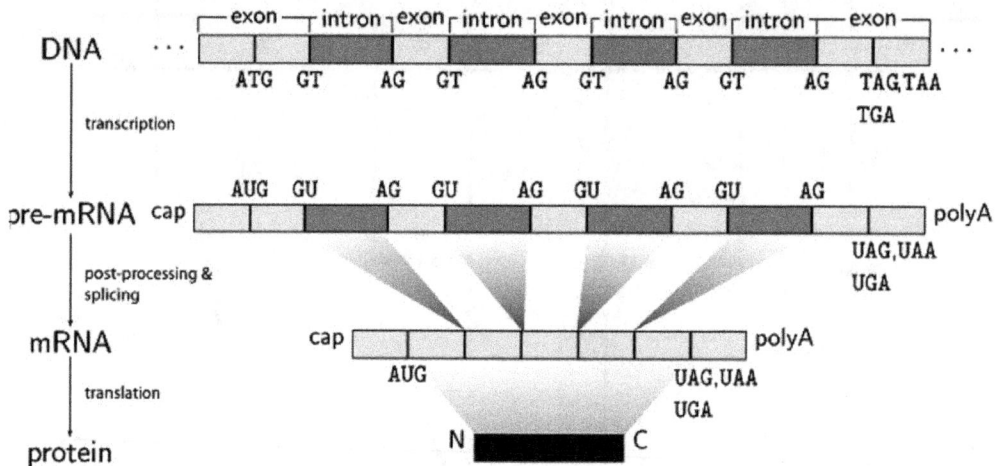

DNA
exon — intron — exon — intron — exon — intron — exon — intron — exon
ATG GT   AG   GT   AG   GT   AG   GT   AG   TAG,TAA TGA

transcription

pre-mRNA cap
AUG GU   AG   GU   AG   GU   AG   GU   AG   polyA
UAG,UAA UGA

post-processing & splicing

mRNA cap polyA
AUG   UAG,UAA UGA

translation

protein N _____ C

Spliceosomes and spliceosomal introns are responsible for splicing out introns and transposable elements, and insuring that the genetic sequences in introns are not translated into proteins. Thus, they regulate gene expression and help guarantee that only designated exons (coding sequences of a gene and its strings of nucleotides) are translated and transcribed (Roy & Gilbert, 2006) thereby producing specific products, organs, functional capacities, and, perhaps, a variety of species.

Introns are directly relevant to the regulation of gene function and expression and RNA processing and have also been highly conserved and preserved often in the same places in the genome, over the course of evolution, be it the genes of Drosophila melanogaster (the fruit fly), Caenorhabditis elegans (nematode), mice, or humans (De Souza et al., 1996; Federov et al., 2002). Again, viruses are a major source of introns.

Viruses can also make copies of genes and then exit the host, only to insert them into the genomes of other species, including other viruses. Therefore, each time a virus is jettisoned into space, it would carry copies of these genes which would then be inserted into the genomes of new hosts when that virus falls upon the surface of another planet. Likewise, these viruses would obtain genes from the denizens of the planets upon which they fell; genes which would be added to the growing viral genetic library. As the number of viruses is innumerable, then so too would be the number of genes and multiple genomes stored collectively within these viral libraries (Joseph 2009b,c,d).

Viral Genetic Libraries

Archae and bacteria are distinguished by the company they keep, i.e. viruses. There are viruses which selectively target Eukaryotes, others which prefer bacteria, and those which are found in association with archae. Then there are viruses which prefer other viruses.

Just as archae and bacteria significantly differ, so too do those viruses which target archae vs those which prefer bacteria (Hendrix, 2004; Pagaling, et al., 2007; Prangishvili et al., 2006). All archaeal viruses discovered so far have DNA genomes (Ackermann, 2007; Prangishvili et al., 2006; Rice et al., 2004) whereas the genomes of bacterial viruses are either RNA or DNA.

Viruses maintain a large reservoir of excess genes. Up to 25% of viruses studied in fact contain up to 3 complete genomes (Goff et al., 2012).

Considerable evidence has been marshaled which demonstrates that viruses are utilized by bacteria as vast storehouses of genes and DNA, which may be transferred from viruses to bacteria, and then back again, depending on environmental and other conditions which impact bacterial needs and requirements. Moreover, viruses, as well as bacteria and archae, can store their genes within the Eukaryotic genome (Conley et al., 1998; López-Sánchez et al., 2005; Romano et al., 2007). Once embedded these genes can be transmitted, in "silent" non-acted form, to subsequent generations and subsequent species, at which point them may be retrieved and/or activated by yet other viruses and transferred to other newly evolving species, or even inserted back into a bacterial or archael genome.

For example, viruses maintain a store-house of genes which code for photosynthesis (Lindell et al., 2004; Sullivan et al., 2005, 2006; Williams et al., 2008). These genes, including those coding for photoadaptation and the conversion of light to energy remain in viral-storage and are only transferred to bacteria under conditions of reduced sunlight and increased environmental stress resulting in nutrient depletion. When these conditions threaten the bacteria with starvation, viruses will transfer the necessary genes to the genome of these starving bacteria.

Once incorporated into the bacterial genome, these genes enhance the cell's photosynthetic machinery so as to obtain the necessary energy and nutrients by capturing additional sunlight (Sullivan et al., 2006). When sufficient sunlight and nutrients become available, these genes are transferred from the bacteria genome back to the virus genome for storage (Lindell et al., 2004; Sullivan et al., 2005, 2006). Viruses and prokaryotes maintain a genetic co-dependency such that genes are commonly transferred back and forth between them on an "as needed basis."

Viruses, or viral-like plasmids, may also be periodically ejected from the archae and bacteria genome as packets of DNA. When these genes are needed, these packets are opened and the necessary genes extracted and inserted into prokaryotic and eukaryotic genomes (Sullivan et al., 2006; Williamson et al., 2008).

Depending on the environment, viruses outnumber bacteria by ratios up to 100 to 1 (Porter et al., 2007; Romancer et al., 2007). These ratios are exactly what might be expected if viruses serve as vast DNA repositories and thus as a source of genes which may be injected into the bacteria genome to be utilized in times of stress or in response to other factors impacting evolution. Stressful conditions would likely be encountered in space, and when journeying to and arriving on different planets where every conceivable environment may be encountered. In fact, virus particles have been found in association with clusters of an extensive array of microfossils similar to methanogens and archae in the Murchison meteorite (Pflug 1984); a meteorite so old it predates the origin of this solar system and may have originated on a planet that orbited the parent star which gave birth to our own (Joseph 2009a).

Viruses are so numerous and come in so many varieties their numbers and genetic storage capacity are essentially infinite. This also means that in total, these viral libraries may contain an infinite number of genes which code for innumerable functions that are held in reserve unless required by the host.

For example, giant double-stranded DNA viruses (such as *Acanthamoeba polyphaga*, Mimivirus), with particle sizes of 0.2 to 0.6 microm and genomes of 300 kbp to 1,200 kbp, maintain vast and complex gene pools (Claverie 2005). These giant double-stranded DNA viruses, such as *Poxviridaem*, also have double-stranded linear DNA genomes which are larger than most bacteria. Therefore, these viruses have incredible genomic capacity and maintain an extensive gene library which was likely obtained via horizontal gene transfer from a host to the virus.

By acting as a genetic storehouse, with the capacity to maintain up to three separate genomes, viruses free up genetic space in the host's genetic machinery which need only maintain those genes necessary for its survival and functional integrity, or the next step in evolutionary change or speciation. Thus, viruses act as gene and genomic conservatories and can increase the gene pool within the genome of a host when necessitated by environmental or other conditions (Sullivan et al., 2006; Zeidner et al., 2005) including those impacting evolution (Joseph 2009b,c,d,e). And they can insert the necessary genes which can produce products which change the environment (which acts on gene selection), and the later retrieve these genes and store them in the viral genetic library.

It is likely that viruses found in association with archae provide a similar genetic-satellite function, orbiting in close proximity and acting as a store house for genes which may be required by archae when confronted with life-threatening challenges.

Thus, viruses serve as genetic luggage and vast storehouses of genes, and within that luggage each viron may possess up to three genomes.

Since viruses are ideally suited to serve as vast genetic libraries, innumerable genes containing vast stores of genetic information would exponentially accumulate within innumerable viral genomes as they were cast upon and the ejected from habitable planets.

Thus, over the eons, and as prokaryotes and viruses in particular, are ejected and transferred to other planets, they would have carried with them vast stores of genetic information; genes that code for innumerable traits and characteristics and which would enable innumerable species to evolve hearts, lungs, bones and brains; genes which promote changes in the environment; genes which respond to changing environmental conditions; genes which enable microbes and other creatures to alter the environment which in turn acts on gene selection thereby giving rise to an evolutionary progression of increasingly complex and intelligent species.

Eventually these microbes, viruses, and their vast genetic libraries, fell to new Earth. And the genetic libraries maintained in the genomes of these viruses and Prokaryotes, made it possible to not only immediately adapt to every conceivable environment, but to biologically modify and terraform new Earth, and in so doing, they began to promote the evolution, metamorphosis, and replication of species which had long ago evolved on other planets.

Viruses Serve the Host: Diseases Are Rare

The defining feature of viruses including retroviruses, is they precisely target specific species and host cells. Further, the viral RNA genome is actually a template for DNA which must have been copied from another source of DNA. However, if that perfect host has not yet evolved, on Earth, then the virus remains dormant. Thus the dormant virus must have obtained its template from a specific extraterrestrial host. This explains why the virus acts purposefully, targeting and inserting its RNA or DNA into specific species after they evolve. By contrast, if there is not a 100% perfect genetic match, errors are introduced thereby harming the host.

Therefore, in some instances, when viruses invade Eukaryotes, they may sicken or even kill the host. However, that is not advantageous to the virus which may also die. Given that viruses exist in vast numbers, and considering that so many viral genes have benefited the host, it appears that it is only relatively rarely that a virus may sicken or kill those they infect. Rather, it could be said that when sickness results, its because these viruses introduced genetic errors into the genetic hardware of the host--perhaps due to a slight mismatch between the inserted gene or viral element, and the genome of the targeted species (Joseph 2009b,c,d).

Viruses store multiple genomes and vast amounts of genetic information which provide no direct benefit to the virus. Viruses instead, serve potential hosts by storing genes which can be selectively transferred to the host depending on need and which provide substantial benefits to the recipient (Lorenc & Makalowski. 2003; Miller et al., 1999; Parseval & Heidmann 2005). For example, some viral genes enhance host cell carbon metabolism, nitrogen fixation, antibiotic resistance, the biosynthesis of vitamin B12, and the creation of heat shock proteins during times of stress (Evans et al., 2009; Sherman & Pauw, 1976; Sullivan et al., 2005: Williams et al.. 2008).

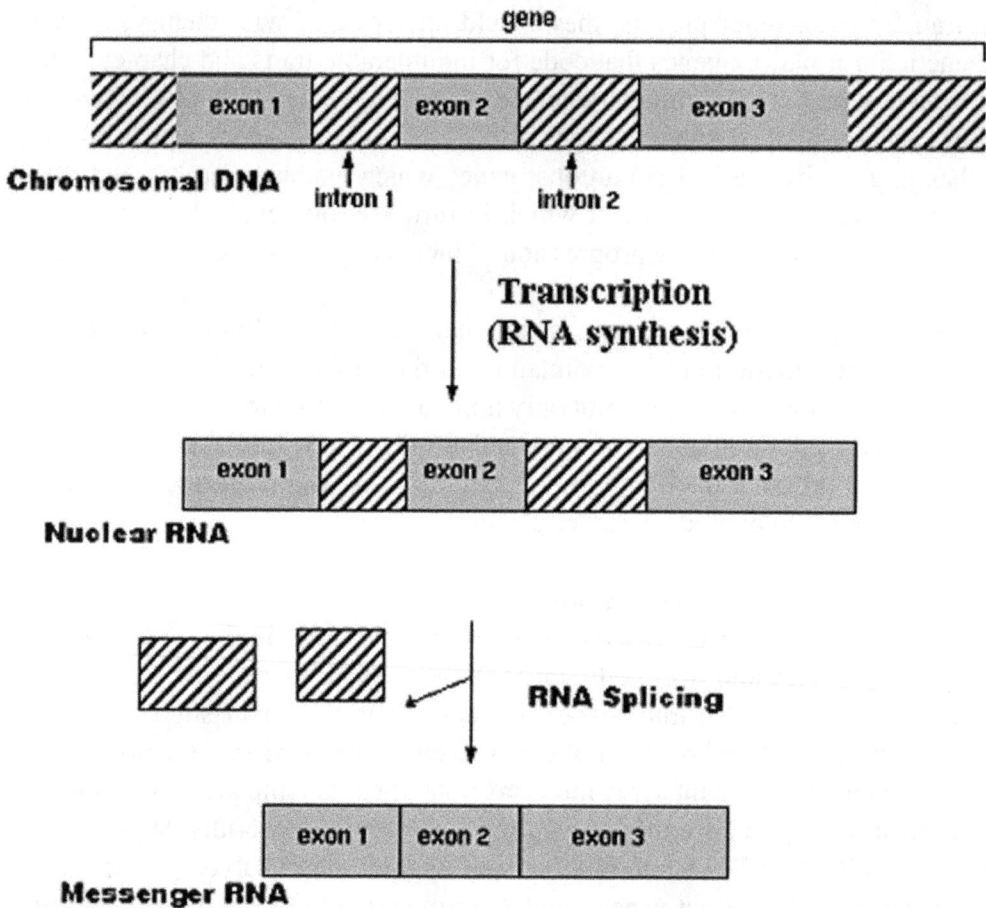

Endogenous retro-viruses (ERVs) are responsible for the generation of proteins involved in the formation of mammalian placenta (Mi et al., 2000; Blond et al., 2000) and provide a protective function allowing for nutrients to pass from mother to fetus while simultaneously protecting the fetus against infection or rejection by the mother's immune system (Ponferrada et al., 2003; Prudhomme et al., 2005). ERVs are also highly expressed in many human fetal tissues including heart, liver, adrenal cortex, kidney, the central nervous system, and human brain (Anderson et al., 2002; Conley et al., 2008; Patzke et al., 2002; Seifarth et al., 2005; Wang-Johanning et al., 2001, 2003).

ERVs are very active in the human genome (Lower et al., 1993; Medstrand & Mager1998). They regulate human gene expression (Jordan et al., 2003; van de Lagemaat et al., 2003) and contribute promoter sequences that can initiate transcription of adjacent human genes (Conley et al., 2008). Endogenous retroviruses, therefore, can alter host gene function and genome structure and thus the evolution of Eukaryotes including humans.

In fact, one of the purposes of viral gene storage is to enable the transfer of specific genes into specific species, just before or after these hosts evolve. The evolution of numerous species including humans, has been shaped by successive waves of viral invasions.

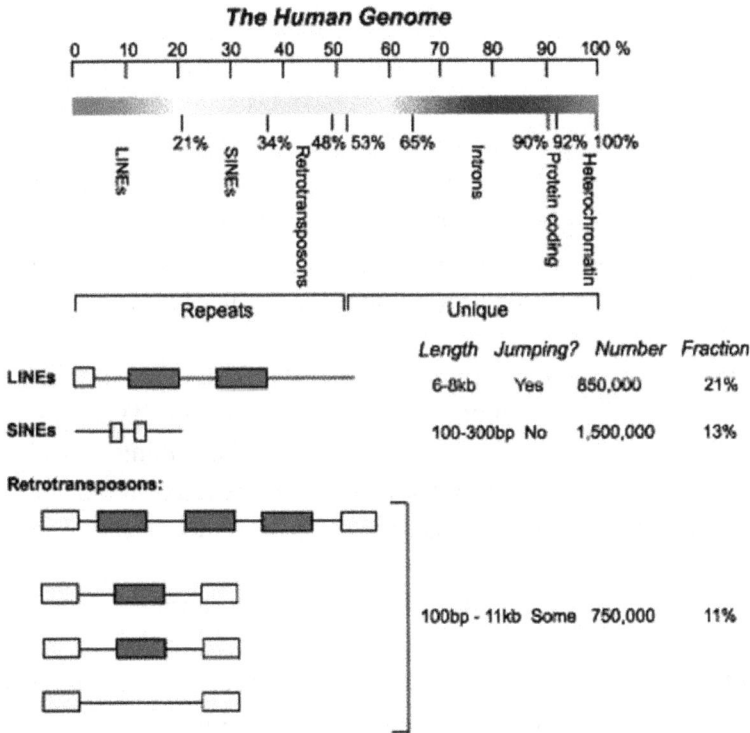

Viruses Target Specific Hosts Before And After They "Evolve"

Flurries of ERV activity, viral invasions, viral gene proliferation, deletions, and extinctions have corresponded to the divergence and metamorphosis of numerous species, and are associated with the Cambrian Explosion 540 million years ago, including the evolution of jawed vertebrates (Agrawal et al. 1998, Kapitonov & Jurka 2005), the subsequent split between fish and tetrapods 450 mya (Volff et al. 2003), the giant leap from teleost fish to amphibians 350 mya (Volff et al. 2001c), then the evolution of reptiles (Hude et al., 2002) and leading up to birds, mammals (Herniou et al., 1998) then primates and humans (Hughes and Coffin 2001; Sverdlov 2000).

Thus we see a leaping, progressive pattern of speciation and primate-human evolution which correspond to viral invasions, increases in viral activity, and the insertion of viral elements. These confluences of viral events exert major influences on the genome, leading to new species including primate-human species which are invaded by yet other viral elements, including waves of group II introns and retrosponsons.

Viral invasions and prokaryotic genetic contributions to the evolution of Eukaryotes leading to humans should not be viewed as agents of chance. Deletions, insertions, bursts of viral activity, and subsequent speciation are under tight regulatory control and involve well coordinated interactions between genes contributed by Prokaryotes and viruses.

Moreover, viruses serve the host, and clearly these genes did not just randomly evolve in viruses for the purpose of benefiting and being injected into the genomes of humans and other species.

Viruses and retroviruses are host specific and precisely target specific species and cells. In order to invade, or for viral genes to become activated, requires the existence of specific species or the evolution of a new host. In other words, each viral *key* had to await the evolution of a specific genetic *lock*. Once that *lock* evolved, the *key* was inserted opening the door to the next stage in evolutionary metamorphosis.

For example, ERV sequences encompass 42.2% of the human genome and almost half of the mammalian genome (Deininger & Batzer 2002; van de Lagemaat et al. 2003); many of which were inserted during key points of evolutionary divergence and speciation. Retroviral sequences encode tens-of-thousands of active promoters and regulate human transcription on a massive scale (Conley et al., 2008).

However, most of the ERV sequences in the human genome are primate-specific, and the entire linage leading from monkeys, apes, hominids to humans, has been continually and selectively targeted by retroviruses which have inserted genes and introns into successive primates species leading to and including humans (IHGSC 2001; López-Sánchez et al., 2005; Medstrand et al., 2002; Romano et al., 2007).

Each stage of evolution and each viral infection event preceded or required that the host first evolve; at which point viruses invaded and inserted genes which interacted with ancient genes which had been donated by Prokaryotes hundreds of millions if not billions of years before. That is, it is only when specific hosts evolve that viral genes or regulatory elements which have been held in abeyance within these viral genetic storehouses, are retrieved, inserted and activated or silenced; all of which is associated with evolutionary divergence and speciation, such as the split between new world and old world monkeys, and the split between hominids and chimpanzees (López-Sánchez et al., 2005; Romano et al., 2007).

For example, 55 mya, with the evolution of monkeys, ERVs formed numerous proviruses which became highly active and increased their activity until the divergence of Old World and New World primates (Lavie et al., 2004). There were more viral invasions following the split between New and Old World monkeys 30 to 35 mya, and new classes of ERVs flourished (Mayer and Meese 2002; Medstrand et al., 1997; Seifarth et al., 1998); each class being due, presumably, to a unique infection event.

During a period of ape-primate proliferation from 15 mya to 6 mya ERVs were again repeatedly mobilized (López-Sánchez et al., 2005), with extensive traces being retained even in the human genome. There followed yet another period of ERV invasion and proliferation 8 to 6 MYA (Barbulescu et al., 1999; Johnson & Coffin 1999; López-Sánchez et al., 2005) corresponding with the divergence between the common ancestors for chimps and humans. And then many of the ERV families became inactive and went extinct (López-Sánchez et al., 2005).

Hundreds of retrogenes also appeared in both the chimpanzee and human lineages after they split from a common ancestor (Chimpanzee Sequencing and Analysis Consortium 2005). Once humans began to evolve there were flurries of viral invasions and ERV activity (Barbulescu et al., 1999; Buzdin et al., 2003; Medstrand and Mager 1998). These viral and ERV invasions have directly contributed to human genetic diversity (Seleme et al. 2006) and the evolution of numerous species of Homo.

The human genome contains 200,000 copies of endogenous retroviruses grouped in three classes (Lander et al. 2001), which have been introduced through at least 31 separate infection events (Belshaw et al. 2005). And yet, there is virtually little or no trace of ERV activity in prosimians--the ancestors of monkeys, apes and humans. Thus, humans, apes, and monkeys have been repeatedly and selectively targeted by viruses which can only infect humans or apes or monkeys. However, numerous viral genes and retro-elements inserted into monkeys did not become active until just before or after the evolution of apes, whereas many of those viral genes inserted into the ape genome did not become active until hominids and then humans had evolved.

Viruses target specific hosts. Some invade bacteria, others amphibians, yet others reptiles, or specific mammalian species. Viruses are selective and are host specific. Before the host "evolves," viruses targeting those hosts remain inactive.

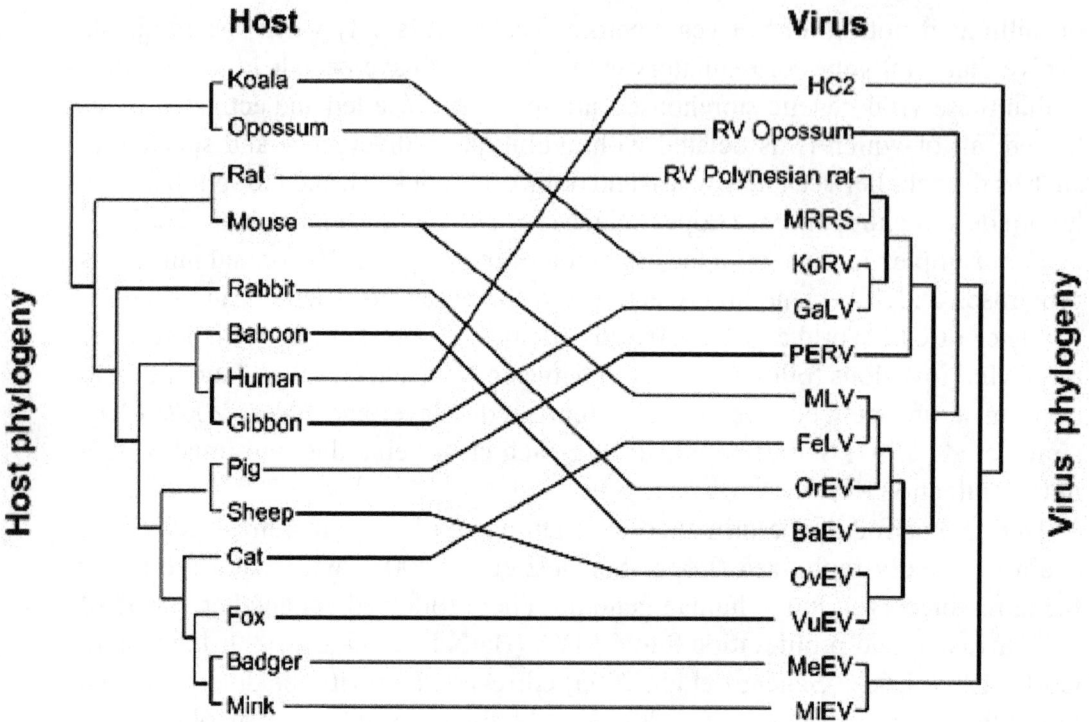

There is a precise *key lock* relationship between a virus and host. The viral RNA or viral DNA genome must perfectly match the DNA genome of the host or there can be no infection. If the match is not completely perfect, then errors are produced which result in disease (Joseph & Wickramasinghe 2010). If the viral-*key* cannot fit into the genetic-*lock*, there is no infection and the virus remains dormant and inactive until a host evolves and becomes available.

Since anatomically "modern" humans appeared around 50,000 years ago, and archaic H. sapiens just a few hundred thousand years ago, then those viruses which selectively target humans could not become active until after humans evolved. Thus, unless these primate-human-host specific viruses just recently fell to Earth in the last 60 million years, then these primate-specific and human-specific viruses must have remained dormant for billions of years before they could become active.

The targeting of specific species can also apply to retroviral genes inserted into the Eukaryotic genome millions and even billions of years ago but which only became active with the evolution of subsequent and specific species. This would explain, for example, why numerous viral genes and retro-elements inserted into monkeys did not become active until just before or after the evolution of apes, whereas many of those viral genes inserted into the ape genome did not become active until hominids and then humans had evolved. That so many of these genes and retro-elements within the human genome do not appear to be active, also indicates that evolution and metamorphosis will not stop with humans.

The fact that the viral genetic key must wait for the evolution of a specific genomic lock, indicates that these genetic keys and locks had evolved long before suitable hosts became available on this planet. That is, since the viral RNA or viral DNA is actually a template for species-specific DNA, and since these viruses exist before these species evolve, then then these viral templates must have been copied from an identical source of host-specific DNA which must have long evolved evolved on another planet. The ease at viral insertion and integration, the fact that the viral gene-host genome are a perfect fit, indicates that the original viral source for this RNA/DNA template of DNA was the DNA of an identical species who long ago lived on other worlds. This explains why host specific viruses exist before the species evolves, why the virus acts purposefully, targeting and inserting its RNA or DNA into specific species or cells, and why errors are introduced if the match is not perfect.

And since there are planets which are not just habitable but billions of years older than this planet and upon which intelligent life may have evolved long before Earth became Earth, this also explains why evolution on this planet will continue into the future.

On the other hand, given the infinite variety of viruses in existence, it may be that most viruses, at least on Earth, never become active, and that the lock-key genetic relationship between host and virus is due to chance. By happenstance the viral key fits the genomic lock of specific cells and species because there are so many viruses.

But then, where did this infinite variety of viruses obtain their genetic libraries and their specific RNA and DNA templates? Why are viral genes beneficial to the host and promote speciation?

Then there is the step wide progression of increasingly intelligent species, and then evolutionary quantum leaps without intermediate forms; all of which arc linked to changes in the environment acting on genes inserted into the Eukaryotic genome by Prokaryotes and the activity of viruses on gene duplication and expression.

Rather than random chance secondary to an infinite variety of viruses which lay in wait, it may be that life on Earth is merely a sample of life's evolutionary possibilities some of which are expressed on planets with their own varied environments. And this genetic potentiality is represented in the genomes of viruses and Prokaryotes, these *genetic seeds of life* which germinated upon falling to Earth.

What is called "evolution" is like metamorphosis and embryogenesis, all of which are under precise genetic regulatory control.

Thus, the data on evolution and metamorphosis presented here, can be interpreted in two ways: **1)** archae, bacteria, and viruses from this world or other planets, have repeatedly inserted genes into the Eukaryotic genome thereby influencing the direction and trajectory of life's evolution on this planet. Or **2)** Extraterrestrial viruses and microbes have repeatedly inserted genes into the Eukaryotic genome, and contain all the genes, genetic elements, and genetic instructions for altering the environment which acts on those genes donated by viruses and Prokaryotes, and that evolution is a form of embryogenesis and metamorphosis: the recreation of creatures who long ago evolved on other worlds.

5. Meteors, Microbes, Viruses: Genetic Seeds of Life Keep Falling to Earth

Interstellar Clouds of Bacteria

Hoyle and Wickramasinghe (1962, 1969), were the first to examine nebular and other diffuse interstellar clouds for signs of biology and life. After first discovering what appeared to be clouds of carbon particles, they later determined, based on spectral analysis, that these carbon grains resembled freeze-dried bacteria (Hoyle et al., 1982; Hoyle, 1982; Hoyle and Wickramasinghe, 1992); i.e. dead bacteria and spores.

As recalled by Hoyle and Wickramasinghe: "When we examined the light scattering properties of freeze-dried bacterial particles (hollow organic grains), a remarkable correspondence with astronomical data emerged. Such a precise correspondence was not found possible for any inorganic, non-biological grain model... A laboratory spectrum of a desiccated bacterium E. Coli, together with a simple modeling procedure, provided an exceedingly close point-by-point match to astronomical data. At this stage we realized that a large fraction of interstellar dust was... spectroscopically... indistinguishable from freeze-dried bacterial material in combination with their degradation products. In our galaxy alone the total mass of this bacterial type material had to be truly enormous."

Hoyle and Wickramasinghe reasoned that "bacteria which have no protective coatings and which are continually exposed to cosmic rays... in the so-called "diffuse" clouds, must be subject to degradation and eventual destruction. The process would be analogous to coalification and graphitization of living material. First, microorganisms expelled from any galactic source into unshielded regions of interstellar space will become deactivated. Then, the deactivated particles will be subject to steadily increasing degradation, ending in a release of free organic molecules and polymers, similar to what astronomers have been discovering since the late 1960's. The ultimate end product will be a transformation of a viable bacterium to a submicron-sized particle" which from the vantage point of Earth, appears to be diffuse clouds of dust which permeate the Milky Way and every galaxy so far observed.

However, these clouds of submicron-sized bacteria may not necessarily be dead, but alive, having shrunk to the size of spores. Typically when bacteria are exposed to life-threatening conditions they shrink in size and form a mineralized core enclosed in heat or cold shock proteins which wrap around and protect them. They will also saturate their DNA with acid soluable proteins which alters the enzymatic and chemical reactivity of their genome making it nearly impermeable to harm. Further, while in the dormant stage a spore has no metabolism and resists cycles of extreme heat and cold, and extreme desiccation. Spores can in fact survive for hundreds of millions of years before returning back to life (Dombrowski, 1963; Vreekand et ak., 2000).

Microbes in Space

It has been demonstrated that bacterial spores and some single celled creatures can survive, with minimal shielding, within the thermosphere and extraterrestrial space environments (Horneck, 1993; Horneck, Bücker, Reitz, 1994; Horneck, Eschweiler, Reitz, Wehner, Willimek, Strauch, 1995; Nicholson, Munakata, Horneck, Melosh, Setlow, 2000). It is also true that isolated spores can be killed if exposed to the full space environment for a mere few seconds. However, if shielded against solar UV radiation, such as if embedded in clay or meteorite powder (artificial meteorites), and presumably if deep within the center of a colony of bacteria with the outer rims consisting of dead bacteria, they can easily survive. The same would be true of microbes and spores just a few inches beneath the surface of a meteor, asteroid, or comets.

In fact, it has been experimentally demonstrated that microbes can survive conditions in space, including ejection from and the crash landing onto a planet, the frigid temperatures and vacuum of an interstellar environment, and the UV rays, cosmic rays, gamma rays, and ionizing radiation they would encounter (Burchell et al. 2004; Burchella et al. 2001; Horneck et al. 2001a.b, Horneck et al. 1994; Mastrapaa et al. 2001; Nicholson et al. 2000).

The fact that microbes of Earth, can survive conditions to which they have never before been exposed, demonstrates that they have inherited these survival mechanisms from ancestors who were exposed to the conditions of space, survived, and then passed on their worthy genes.

Only an ancestry which leads to space environment can explain why microbes born on this planet are already pre-adapted for journeying through space, living in space, and not just surviving but flourishing in radioactive environments where they are continually exposed to radiation by ions similar to what might be encountered in a nebular cloud.

Radiation Resistant Microbes

In 1958, physicists discovered clouds of bacteria, ranging from two million bacteria per cm3 and over 1 billion per quart, thriving in pools of radioactive waste directly exposed to ionizing radiation and radiation levels millions of times greater than could have ever before been experienced on this planet (Nasim and James, 1978). The world's first artificial nuclear reactor was not even built until 1942. Prior to 1945, poisonous pools of radioactive waste did not even exist on Earth. There is no way these microbes or their immediate ancestors could have been exposed to conditions even remotely as radioactive; which also means that this could not be an example of natural selection survival of the fittest on Earth where survivors passed on their genes.

And yet, over a dozen different species of microbe have inherited the genes which enable them to survive conditions which for the previous 4.6 billion years could have only been experienced in space. These radiation-loving microbes include Deinococcus radiodurans, D. proteolyticus, D. radiopugnans, D. radiophilus, D. grandis, D. indicus, D. frigens, D. saxicola, D. marmola, D. geothermalis, D. murrayi, and so on.

Simple Eukaryotes including lichens, fungi and algae can also survive exposure to massive UV and cosmic radiation and the vacuum of space (Sancho et al. 2005). Likewise, protozoon Acanthamoeba are radiation resistant (Hijnen et al., 2006). Many of these species, including Bacteria can rebuild their genomes even if shattered by radiation (Lovett 2006; Scheifele and Boeke 2008). Thus, single celled Eukaryotes also have a genetic ancestry inherited from those preadapted to journeying through space, and raising the possibility this domain may have also been born in space. This also means that single celled Eukaryotes may have also arrived on Earth from space, along with archae, bacteria, and viruses (Joseph 2009a,b); and this would explain the discovery of microfosils resembling yeast cells and fungi dated to 3.8 bya (Pflug 1978).

Radiation Resistant Viruses

Many types of Viruses are also radiation resistant (Fekete et al., 2005; Gibbs et al., 1978; Hijnen et al., 2006; Jung et al., 2009). Moreover, freezing temperatures will increase the radiation resistance of various species of Virus (Jung et al., 2009). Thus, Viruses, Bacteria, Bacterial spores, protozoon, lichens, fungi and algae, are radiation resistant (Hijnen et al., 2006; Lovett 2006; Sancho et al. 2005) and they could have only acquired this resistance through the inheritance of genes passed on through natural selection following exposure in a radioactive environment other than Earth.

Like Bacteria, Viruses have been shown to survive simulated extraterrestrial conditions (Fekete et al., 2004; Walker, 1970). For example, in one set of experiments, bacteriophage T7 and isolated bacteriophage T7 DNA were exposed to space conditions in the international space station including vacuum and UV radiation and temperatures of 0°C. It was determined that DNA lesions will accumulate but the amount of damage is inversely proportional to the thickness of shielding and layers (Fekete et al., 2005). With increased shielding, such as might be expected if encased in a meteor, asteroid, or comet, the damage is minimal. Further, following simulated space conditions including prolonged radiation, up to 60% of T7 phages remained active and were able to infect Bacterial host cells, and those phages suffering damage were able to fully recover (Fekete et al., 2004).

Viruses, including those with double-stranded DNA genomes have also been shown to survive in the most extreme of environments (Pagaling, et al., 2007; Prangishvili et al., 2006; Rice et al., 2004; Romancer et al., 2007; Walker,1970) including extremely acidic hot springs with temperatures up to 93°C, and pH 4.5 (Häring et al., 2005; Rice et al., 2001, 2004), in hypersaline water at saturation (Porter et al., 2007), a well as in deserts, soda lakes, deep sea thermal vents, and under incredible hydrostatic pressures (Romancer et al., 2007). Likewise, wild type filamentous phage M13 retained their nucleic acid integrity and protein structure despite high pressure and even simulated silicification (Hall et al., 2003).

Archae viruses, and other prokaryotic extremophiles, are able to flourish under these same life-neutralizing conditions (Pagaling, et al., 2007; Porter et al., 2007; Prangishvili et al., 2006; Rice et al., 2004; Romancer et al., 2007).

Microbes, viruses, and some single celled eukaryotes from Earth are preadapted to surviving conditions which they have not encountered on this planet. Therefore, they must have inherited these genes which made survival in space possible, which in turn could only have been acquired from ancestors which had survived when exposed to or born in a space environment (Joseph 2009a). Because microbes and viruses can survive the space environment and a journey through space with minimal shielding, they are the perfect vehicle for spreading the genetic seeds of life throughout the cosmos.

Sojourners from the Stars

How could these life forms, these "seeds of life" journey from a nebular cloud, travel through space, and then land and take root on a planet, such as Earth?

Although the full spectrum of UV rays is deadly against bacteria and spores, the likelihood of a direct hit, even if unprotected while traveling through space is unlikely. Estimates are that a spore may journey for up to a million years in space before it may be struck by a UV ray (Horneck et al., 2002). Thus, given trillions upon trillions of spores drifting through space, it is likely a significant percentage might survive; even more so if they were attached to cosmic dust. And, it has been estimated that a microbe drifting through space, could traverse a distance of 4.37 light years (1.34 parsecs) in just 9,000 years (Arrhenius, 2009), which is the distance from Alpha Centauri to Earth, and approximately 90,000 years to traverse the 400 light year distance from the Helix Nebula to this planet.

Vreeland and colleagues (2000) discovered Bacteria spores which had been embedded in salt crystals buried 569 meters beneath the earth, and dating back 250 million years. These dormant spores, Bacillus permians, also came back to life and began to multiply. In 1963, H. Dombrowski also brought back to life Bacteria which had been embedded in salt deposits from the Middle Devonian, the Silurian, and the Precambrian (Dombrowski, 1963). Some of these Bacterial spores managed to survive for over 600 million years.

Thus it can be predicted that Bacteria spores could journey across the cosmos for up to 600 million years. And once they fell upon a suitable planet, they could go forth and multiply. However, in so doing, they, and any viruses accompany them could also infect those they contact.

Comets: Wombs of Life Delivery

If microbes and viruses swarm throughout the cosmos, then comets sweeping through nebular and other stellar clouds brimming with life, could vacuum up trillions of microbes with those coming to reside deep within the comet's interior thriving and multiplying. Hence, when these comets near any of the billions of suns within the Milky Way or any other galaxy, clouds of bacteria would be released into the comets tails. When any moon or planet, including Earth comes into contact with the tail of a comet they become contaminated with life; and this is how life may have originated on this planet.

It has been determined that comets eject particles at a rate of a tens of thousands of tons a day, sometimes in great bursts as was observed for Halley's Comet on March 30-31, 1986. In fact, infrared observations and analysis of the infrared emission spectrum of dust from Halley's Comet obtained by Dayal Wickramasinghe and David Allen in March 1986 matched precisely the laboratory spectrum of dessicated bacteria. These discoveries indicate that a large percentage of cometary particles, just like interstellar particles, are spectroscopically identical to bacteria. The particles therefore may be bacteria and spores of bacteria which are being sprayed along with their degradation products into the space medium, and onto any nearby planets and moons.

Red Giants, Solar Winds and Biomass Blown Into Space

This Earth is estimated to have been formed and a member of this solar system 4.6 billion years ago. However, in this galaxy alone, there are stars, and thus planets, which are over 13 billion years in age. And then there are the innumerable stars which became red giants and then underwent supernova billions of years before Earth became Earth. If any of the ancient planets belonging to these ancient solar system were home to life, the smallest of these life forms could have easily been dispersed into space, into nebular clouds, swept up by comets, and deposited on other worlds, including those, such as this Earth, which had not yet been created.

Any planet with oceans, atmosphere, and surface dwelling organisms will inevitably seed surrounding moons and planets with microbes and possibly eukaryotic life (Joseph 2000, 2009a). If they survive once they arrive on these other planets and moons would be determined by multiple factors. Microbial organisms from a single planet may even come to be distributed on a galaxy-wide scale, some of which would come into contact and exchange DNA with microbes expelled from or living on these other worlds (Joseph 2000, 2009b).

The mechanisms of dispersal are many and include A) Solar winds, B) Bolide Impact, C) Comets, D) Ejection of living planets prior to supernova which are then captured by a newly forming solar system, E) Galactic collisions and following the exchange of stars between galaxies.

A living planet with an atmosphere, orbiting within the habitable zone of a sun-like star, will be subjected to that star's solar winds which, if sufficiently powerful, could blow living biomass into space. Earth's powerful magnetic field usually protects the planet from these winds. However, these solar winds buffeting Earth periodically increase significantly in strength, overcoming the protective magnetic field, and can eject air-born microbes into space and distribute them not just to neighboring planets, but outside the solar system where they may come to contaminate collections of "Oort cloud" stellar objects and passing comets. After 9,000 years some of the survivors could land on a planet in Alpha Centauri which is 4.37 light years distant, and in a few million years come into contact with tens of thousands of planets throughout this galaxy.

There are 29 other stars within 12 ly of Earth and which are similar enough to the sun to possibly sustain life on orbiting planets. A single microbe could be ejected from any one of them and reach our planet in 25,000 years. Even stars 250 ly (and 515,000 years) away could deliver Viruses and living spores to Earth and other planets. In fact, only a single microbe needs to survive to repopulate and cover a suitable planet with microbial life (Joseph 2000, 2009a).

The same process could take place on hundreds of billions of planets in the Milky Way and other galaxies, such that these "seeds of life" could swarm throughout the cosmos, contaminating every habitable planet with new life.

Consider Earth as an illustrative example. Distinct species of over 1,8000 different types of bacteria and other microbes thrive and flourish within the troposphere, the first layer of the Earth's atmosphere (Brodie et al. 2007). Microorganisms and spores have been recovered at heights of 40 km (Soffen 1965), 61 km (Wainwright et al., 2010) and up to 77-km (Imshenetsky, 1978). These include Mycobacterium and Micrococcus, and fungi Aspergillus niger, Circinella muscae, and Penicillium notatum 77-km above the surface of Earth (Imshenetsky, 1978).

Air is an ideal transport mechanism and serves as a major pathway for the dispersal of Bacteria, Virus particles, algae, protozoa, lichens, and fungi including those which dwell in soil and water. Moreover, due to tropical storms, monsoons, and even seasonal upwellings of columns of air (Randel et al., 1998), microbes, spores, fungi, along with water, methane, and other gases may be transported to the stratosphere.

During the Monsoon season of tropical storms, microbes circulating in the troposphere and even those residing on the surfacc of the oceans and earth would easily be lofted into the stratosphere which sits just above the troposphere and extends from 8 km (5 miles) to 50 km (31 miles) in altitude. The monsoon is one of the most powerful atmospheric circulation systems on Earth and commonly funnels air, dust, water, gases, and pollutants from the lower layers of the atmosphere to deep within the stratosphere where they stay aloft and circulate the globe for years (Randel et al., 2010). Further, there is a normal pattern of seasonal upwelling where water, methane, and other gases are transported to the stratosphere in the subtropics and polar regions, by semiannual oscillations in weather, climate, and other factors related to the changing seasons (Randel et al., 1998). Thus, it can be readily assumed that microbes not only flourish in the troposphere, but are commonly lofted into the stratosphere (Wainwright et al., 2010).

Normally, such creatures might be too heavy to be ejected into space. However, when the CME struck on Sept. 24, 1998, the pressure of the solar wind jumped to 10 nanopascals whereas normally the pressure is around 2 or 3 nanopascals. Naturally airborne microbes living in the upper atmosphere would have also been cast into space.

Specifically, as detected and measured by NASA's Ultraviolet Imager aboard the Polar spacecraft, between September 22- 25, 1998, a series of coronal mass ejections (CME) and a powerful solar solar wind created a shock wave which struck Earth's magnetosphere and the polar regions with so much force that oxygen, helium, hydrogen, and other gases were ripped from the Earth's upper atmosphere and ejected into space (Moore and Horwitz, 1998). Therefore, it could be predicted that tons of bacteria, archae, viruses, and single celled eukaryotes were also cast into space.

In 1859, the Earth was struck by a "solar superstorm" which lasted from August 28 until September 2 (Tsurutani et al., 2003). It has been determined that a CME takes three to four days to reach Earth. However, in this instance, an earlier burst cleared a path, and the "solar superstorm" which followed in its wake struck the Earth in less than 18 hours, and with such force that it caused a world-wide failure of telegraph systems. The atmosphere was directly impacted with such a blow that much of the planet was enveloped in shimmering sheets of greens, reds, and blues which were so brilliantly bright that night became day and even the darkest shadows of evening were illuminated with dazzling lights. Certainly the pressure of the solar wind was well above 10 nanopascals (Tsurutani et al., 2003) and quantities of atmospheric gasses, including airborne microbes, and perhaps other creatures, would have been blown into space.

The frequency and fluctuations in the power and force of CMEs and the solar winds, over the course of Earth's history, is unknown. However, an analysis of ice cores indicates that CMEs of an intensity similar to or greater than that of 1859, occur at least once per 500 years (Odenwald and Green, 2008; Tsurutani et al., 2003).

Therefore, it can be predicted that these solar events are not uncommon but must take place on innumerable habitable planets orbiting within a habitable distance from their sun.

Once lofted into space, microbes and spores might easily survive. Microbes and spores are so small that even when bombarded with photons and deadly gamma and UV rays the likelihood they would be struck is infinitesimally minute. Even if struck, the radiation dose would be minimal and the damage might not be fatal. If the organism's DNA is damaged, it can be rebuilt when the spore germinates. Some species of microbe, such as Deinococcus radiodurans, can quickly rebuild their genome even if shattered by UV or gamma rays (Lovett, 2006), and the same is true of yeast (Scheifele and Boeke 2008).

Many species of microbe can withstand X-rays and atomic radiation, and are radiation resistant. Therefore, even microbes which are lofted into space by powerful solar winds would likely survive unscathed.

However, not just microbes, but dust and debris would also be cast into space by powerful solar winds. Innumerable microbes could hitch a ride and attach themselves to these particles. Dust particles are too small to be hit by photons but are the perfect size to reach escape velocity. Although many microbes would die, most might easily survive the conditions of space, protected from radiation by dust and debris (Clayton, 2002; Flanner et al., 1980; Herbst and Klemperer 1973; Nishi et al., 1991; Prasad, and Tarafdar 1983). These microbes need only form spores to survive under these conditions.

Bacteria can in fact sense a life-threatening event even before it occurs and will undergo a sequence of developmental changes to protect itself from death--often with the aid of a Virus which immediately transfers spore-triggering genes into the Bacterial genome. These microbes immediately begin to transform, secreting protective gels, shrinking to the size of spores, and generating heat and cold-shock proteins which wrap around and protect them. The resulting spore then becomes dormant (Marquis and Shin 2006; Setlow and Setlow 1995). Thus, spores could form before they were ejected from the planet, thereby greatly enhancing the likelihood of their survival.

If the solar wind was produced by a dying star during the red giant phase, interstellar space would be thick with dust which would begin to accumulate in a growing nebular cloud. Once these space-journeying microbes become part of the growing nebular debris field, those deposited in the inner layers of the cloud would be protected against deadly gamma and cosmic rays.

Moreover, many species of microbe have inherited the ability to survive a violent hypervelocity impact and extreme acceleration and ejection into space including extreme shock pressures of 100 GPa; the frigid temperatures and vacuum of an interstellar environment; the UV rays, cosmic rays, gamma rays, and ionizing radiation they would encounter; and the descent through the atmosphere and the landing onto the surface of a planet (Burchell et al. 2004; Burchella et al. 2001; Horneck et al. 2001a.b, Horneck et al. 1994; Mastrapaa et al. 2001; Nicholson et al. 2000, 2004; Mitchell and Ellis 1971). Therefore, powerful solar winds, or any phenomenon, such as meteor impact, could cast trillions of microbes into space, and many of them would survive.

Additional support for these scenarios are provided by the discoveries of Hoyle and Wickramasinghe (2000) who discovered clouds of cosmic dust comprised of desiccated bacteria (Allen and Wickramasinghe, 1981). Thus it appears that clouds of bacteria may permeate space. Additional evidence of these bacterial clouds was provided by telescope observations of the galactic centre infrared source GC-IRS7 (Allen and Wickramasinghe, 1981) and of fluorescence phenomenon and extended red emissions in nebula and extragalactic systems, i.e. biological chromophores (pigments), including chloroplasts and phytochrome. It is probable that these bacteria were either born in these clouds, or they are the descendants of those lofted into space by powerful solar winds along with dust and perhaps even surface debris from the various planets effected.

Comets and Meteors: Invaders From Space

Hundreds of millions of meteors, comets, asteroids, and planetary debris collided with this planet during the first 700,000 years of its history, long after life had become established on Earth. Microbes and viruses were contained within that debris. When these subsequent arrivals fell upon Earth it can be predicted they also engaged in HGT with the descendants of those who were among the first to take root on this planet.

Meteors and other debris continued to strike this planet after the initial 700,000 years of extraterrestrial bombardment, albeit with diminishing frequency continuing to the present (Cellino & Dell'Oro 2009; Elewa & Joseph 2009; Napier 2009; Radice 2009). One need only gaze at the meteor cratered surface of the moon to recognize we live in the midsts of a cosmic shooting gallery. It has been estimated that approximately 84,000 meteors weighing up to 1 kilogram hit Earth each year (Bland et al. 1996), whereas larger meteors, ranging from the size of a car to a house, strike the upper atmosphere on a weekly basis.

The Chelyabinsk Meteor of 2013

On February 15, 2013, a meteor which weighed 11,000 metric tones exploded about 15 miles (24 kilometers) over the Russian city of Chelyabinsk. Upon fragmentation, bus-sized and house-sized fragments remained intact until striking the ground with such force they shattered windows and injured over 1,000 people. However, before and after the meteor exploded and as pieces fell, it also shed an enormous cloud of small rocks and dust, creating a plume consisting of hundreds of tons of material which circulated and lingered in the Earth's upper atmosphere, the stratosphere, for over 3 months. If that huge meteor harbored life within its interior, it can be predicted that trillions of microbes and viruses were sprayed into the atmosphere as the meteor disintegrated and before it exploded, and that trillions more survived within the bus-sized debris which fell to Earth.

It has been well established that microbes can survive conditions in space, including ejection from and the crash landing onto a planet, the frigid temperatures and vacuum of an interstellar environment, and the UV rays, cosmic rays, gamma rays, and ionizing radiation they would encounter (Burchell et al. 2004; Burchella et al. 2001; Horneck et al. 2001a.b, Horneck et al. 1994; Mastrapaa et al. 2001; Nicholson et al. 2000). However, if protected by dust, or buried within the depths of a meteor, and as the overlying thickness increases beyond 30 cm, the dose rate and lethal effects of heavy ions, including secondary radiation, depreciates significantly. Even after 25 million years in space, a substantial number of spores would survive if shielded by 2 meters of meteorite (Horneck et al., 2002). The Chelyabinsk meteor weighed 11,000 metric tones. Moreover, spores have been discovered to survive harsh conditions and to return back to life even after the elapse of 250 million (Vreeland et al. 2000) to 600 million years (Dombrowski 1963). Experiments have shown that microbes can easily survive the shock waves of a violent impact even following high atmospheric explosions (Szewczyk et al., 2005) and reentry speeds of 9700 km h-1 (McLean et al., 2006). They can also survive the descent to the surface of a planet (Burchella et al. 2001; Burchell et al. 2004; Mastrapaa et al. 2001; Horneck et al. 2002).

Further, bacteria and other microbes form colonies which serve protective functions. If cast into space, deep inside a mound of earth and stone, those on the outer layers of the colony, if killed, create a protective crust, blocking out and protecting those in the inner layers from radiation or other hazards associated with space travel. Be they buried within rock, ice, or some other stellar material, and regardless of the depth, colonies of living microbes would provide their own protection from the hazards of space, with those who die ensuring the survival of those at the center of the colony. In fact, the fossilized remnants of bacterial colonies have been discovered in a number of meteors, including the Orgeuil, Murchison, and Efremovka meteorites (Hoover 2011; Pflug 1984; Zhmur & Gerasimenko 1999).

When meteors strike the atmosphere they are subjected to extremely high temperatures for only a few seconds. If of sufficient size, the interior of the meteor will stay relatively cool, with the surface material acting as a heat shield; and this would be especially true of bus- and house-sized fragments which fall to Earth intact. The heat does not effect the material uniformly and the interior may never be heated above 100°C (Horneck et al., 2002) thereby leaving thermophiles unscathed. In fact, spores can survive post shock temperatures of over 250°C.

Consider, again, the huge meteor which exploded over the Russian city of Chelyabinsk on February 15, 2013. The meteor weighed 11,000 metric tones and after it exploded and began to disintegrate, tens of thousands of baseball and basketball sized pieces including those as large as a house remained intact. Then there were the tons of material which remained aloft in the atmosphere for months. Thus, if this debris harbored microbes and viruses, trillions of these species would have survived. The same would be true of all large and small meteors which strike the Earth; which means this planet has likely been repeatedly subject to extraterrestrial invasions by microbes and viruses.

Alien Viruses, Genetics, Cambrian Explosion: Humans

Microbes in the Upper Atmosphere

Every day about 100 tons of meteoroid and extraterrestrial material falls to Earth, much of it dust which circulates for months in the stratosphere and then gently waft down until finally falling upon the surface of this planet. If microbes, spores, and viruses were attached, they too would drift down.

As demonstrated by impact studies, small particles gently decelerate when they strike the upper atmosphere and then slowly fall to Earth and any microbes attached to these particles would survive (Anders 1989). They would not just survive, but begin to multiply.

In fact, biological-cell-like particles have been discovered in the upper atmosphere, falling toward Earth, and which appear to be extraterrestrial in origin (Imshenetsky, 1978; Greene et al, 1962, 1965; Soffen 1965). In a report by G. A. Soffen of NASA's Jet Propulsion Laboratory (Soffen 1965), microorganisms and spores were collected using small biological balloon samplers, at heights between 125,000 to 130,000 feet (over 23 miles / 40 KM); at the same height where the Chelyabinsk meteorite exploded on February 15, 2013.

Microorganisms and spores have subsequently been recovered at 61 km (Wainwright et al., 2010, 2013) and up to 77-km above Earth (Imshenetsky, 1978); which is the height at which the Chelyabinsk meteorite began to disintegrate, shedding a plume of dust and debris in its wake. Those microbes recovered at heights of 77-km include Mycobacterium and Micrococcus, and fungi Aspergillus niger, Circinella muscae, and Penicillium notatum (Imshenetsky, 1978).

Viruses From Space

Numerous species of virus can survive massive amounts of radiation (Fekete et al., 2005; Gibbs et al., 1978; Hijnen et al., 2006; Jung et al., 2009), as well as freezing temperatures which actually increase the radiation resistance of various species of viruses (Jung et al., 2009).

Viruses, like bacteria, can also survive simulated extraterrestrial conditions (Fekete et al., 2004; Walker, 1970) and even direct exposure to space conditions including vacuum and UV radiation and temperatures of 0°C, with survival increasing if provided minimal shielding (Fekete et al., 2005). With increased shielding, such as might be expected if encased in a meteor, asteroid, or comet, the damage is minimal. Further, following simulated space conditions including prolonged radiation, up to 60% of viruses not only survive and remained active but were able to infect bacterial host cells and insert their genes (Fekete et al., 2004).

Viruses can survive in the most extreme of environments (Pagaling, et al., 2007; Hall et al., 2003; Prangishvili et al., 2006; Rice et al., 2004; Romancer et al., 2007) including extremely acidic hot springs with temperatures up to 93°C, and pH 4.5, in hypersaline water at saturation, in deserts, soda lakes, deep sea thermal vents, under incredible hydrostatic pressures, and despite simulated silicification. Likewise, it is well established that archae and bacterial extremophiles can survive extremely hazardous, radioactive, toxic, or extremely hot or cold conditions.

Viral Invasions from the Stars

A continual influx of extraterrestrial bacteria, archae, and especially viruses, could account, in part, for the successive invasions of Prokaryote genes and in particular, viral genes throughout the course of evolutionary development leading increasingly intelligent Eukaryotic species.

It was during the Cambrian Explosion that all manner of species suddenly evolved hearts, bones, brains, and eyes, and it was also at the outset of this explosion of life, that flurries of various viruses invaded numerous hosts. The insertion of viral genes, around 540 mya, completely reorganized targeted genomes, leading to the divergence and metamorphosis of numerous species, including the evolution of jawed vertebrates (Agrawal et al. 1998, Kapitonov & Jurka 2005).

Yet another viral invasion and flurries of ERV activity took place during the split between fish and tetrapods 450 mya (Volff et al. 2003), and yet again 350 mya, corresponding with the giant leap from teleost fish to amphibians (Volff et al. 2001c), and again around 320 mya with the metamorphosis of reptiles from amphibians (Hude et al., 2002). There were yet more viral invasions and increases in ERV activity which corresponded to the metamorphosis of mammals from reptiles (Piskurek & Okada 2007), and then again with prosimian primates, and the first monkeys, then apes, followed by hominids then humans (Herniou et al., 1998; Hughes & Coffin 2001; IHGSC 2001; López-Sánchez et al., 2005; Medstrand et al., 2002; Romano et al., 2007).

These invasions and suddenly increases in ERV activity do not prove that these viruses originated from space when meteors continued to strike the planet. Nevertheless, these data, coupled with numerous studies already reviewed, support the proposition that alien, extraterrestrial microbes and viruses buried beneath the surface of innumerable meteors and asteroids, or even merely attached to cosmic dust, have continued to fall upon the surface of this planet and were the first denizens of this Earth--which accounts for discoveries of biological activity in this planet's oldest rocks. Further, it can be deduced that some if not much of the cosmic debris which continued to be deposited on Earth from space after the initial bombardments which lasted 700 million years, probably also harbored life.

Once any subsequent extraterrestrial microbes and viruses arrived on Earth, they would transfer genes to the microbes, viruses, and Eukaryotes already residing on this planet and which had also already engaged in HGT. And just as Prokaryote and especially viral genes have influenced the trajectory of evolution and metamorphosis over 4 billion years ago, it can be predicted that later arriving Prokaryotes and viruses did the same.

The Future is Evolution

If viruses and microbes haven fallen and continue to fall to Earth, it can be predicted they have or will transfer genes to various species of Eukaryotes, including humans and those species who have not yet evolved. It can also be predicted, based on the number of inactive ERVs and silent viral Prokaryote genes within the Eukaryotic including the human genome, that evolution is an ongoing process which will continue billions of years into the future.

"Evolution," metamorphosis, does not stop with modern humans. If humans do not self-destruct and become extinct, it can be predicted that this process of metamorphosis will lead to increasingly intelligent and technologically advanced species who in turn will be targeted by viruses whose DNA/RNA genomes are templates fashioned and obtained from life which evolved on worlds much older than our own.

6. Extinction, Metamorphosis, Evolutionary Apoptosis, Genetically Programmed Mass Death

At the level of DNA, all Earthly-life forms are linked (Joseph 2000a). Further, genes are continually horizontally transferred between Prokaryotes, between Prokaryotes and Eukaryotes, and between Eukaryotes and from viruses to viruses and from viruses to Prokaryotes and Eukaryotes (Aravind et al, 1998; Forterre 2006; Hotopp et al. 2007; Iyer et al. 2006; Koonin 2009; Martin et., al. 2002; Nelson et al. 1999; Nikoh et al. 2008; Zambryski et al 1989). The exchange of genes is not random or due to chance, but under genetic regulatory control--much like the interactional precision of the inner workings of a digital clock. However, in this case, the "clock" is alive and can be likened to a supra-DNA-organism, or a giant all encompassing cellular entity whose component parts which include all species, interact according to biological laws and principles to maintain the health of the organism and to promote not just its survival, but its continual growth. And that growth may involve the shedding of "cells" and components parts which are no longer useful to the supra-organism. That is, not just evolution, but the extinction of species is under genetic control (Joseph 2000a, 2009e).

Genes, cells, and entire species undergo evolutionary apoptosis and are continually pruned from the tree of life. Programmed death is essential to life and evolution, and genetically programmed evolutionary apoptosis is one of the many causes of species death and extinction (Elewa & Joseph 2009; Joseph, 2009e).

Evolution and extinction can be likened to embryogenesis and metamorphosis as all involve the selectively turning on and off of specific genes and nucleotide sequences, the shedding of cells and tissues which are replaced, and dramatic alterations in the organs and skeletal muscular system (Joseph 1996, 2000a,b, 2009e).

"Evolution" is under genetic regulatory control, in coordination with the biological activity of viruses and single celled prokaryotes (archae, bacteria, Cyanobacteria), their donated and horizontally transferred genes (including those which gave rise to mitochondria), and the genetically engineered environment. The interaction between the environment and genetic activity, the secretion of chemicals, enzymes and gasses such as oxygen and calcium, regulates the emergence of new species, and the elimination of yet others--a form of evolutionary-apoptosis (Joseph, 2009e).

Genes (via the organisms they reside it) act on the environment, and the changing environment acts on gene selection, activating specific genes, silencing others, and giving rise to new species which emerge from the old, with entire populations of genes, cells, tissues, and species proliferating and others dying out. Like programmed cell death, extinction is often intrinsic to and necessary for the development, evolution, and metamorphosis of increasingly complex species (Elewa & Joseph 2009; Joseph, 2009e).

Extinction

In the history of our planet there have been at least five major mass extinctions, and a number of minor extinctions (Elewa 2008; Raup 1992; Raup & Sepkoski, 1982). These include the Ordovician Mass Extinction, the Devonian, Permian, Triassic, and Cretaceous Mass Extinction.

In addition to the big five, some scientists believe there have been additional major mass extinctions, including as many as 4 extinctions during Cambrian era. According to Joseph (2009e) these additional extinction events include the Paleoproterozoic (2.3 to 1.8 bya), the Sturtian (725 mya to 670 mya), the Marinoan/ Gaskiers (640 to 580 mya), and the Ediacaran extinctions (540 mya).

Many scientists also believe we are now experiencing a sixth mass extinction, which is driven by Homo sapiens (Crutzen & Stoermer 2000; Elewa 2008a; Jones 2009; Ruddimann 2005; Steffen et al. 2007). There is considerable evidence that extinction has been accelerating over the last 500 years; and with the advent of weapons of mass destruction, and industrial poisons, pharmaceuticals, and other wastes which are dumped into the oceans and atmosphere (Levy and Sidel 2009; McKee 2009, Tonn 2009); it could be said the human race is flirting with self-destruction and may trigger a world-wide mass extinction which could wipe humanity from the face of the Earth (Jones 2009). On the other hand, it may be that instead of extinction, the substances pumped into the environment by humans will and is acting on gene selection, thereby preparing not for the next extinction event, but for the next stage of evolutionary metamorphosis.

Sometimes, numerous species may die off simultaneously, resulting in a mass extinction, or a few individual species may die out in isolation leaving the vast majority unscathed. Extinction is so common that it can be considered an integral and perhaps an essential feature of life on Earth (Bradshaw & Brook 2009; Elewa 2008; Ward 2009). However, the vast majority of these extinction events may have been of just a few species. Mass extinctions may account for less than 5% of all species which have become extinct (Erwin 2001) .

Evolutionary Apoptosis and the Gaia, Medea, Cronus Hypotheses

There are several competing explanations for why certain species are eradicated and for what causes mass extinctions (Elewa & Joseph 2009). In what could be described as the battle of the metaphors (Glickson 2009): the Medea hypothesis (Ward 2009) and the Cronus hypothesis (Bradshaw & Brook 2009) focus on biological self-destruction, whereas the Gaia hypothesis (Lovelock & Margoulis, 1974) presents the Earth and life itself as an interacting organism which has the potential to live and grow, or die. Ward (2009) theorizes the mass extinctions of species is in part driven by life itself, and the interaction of biological processes forces species self-destruction. Bradshaw and Brook (2009) see mass extinctions as a natural consequence of the "ebb and flow of life on Earth along a thermodynamic spectrum" and that "the causes of extinction can be thought of as equivalent to the different processes that lead to individual deaths within a population" albeit in the context of Darwinian natural selection.

The individual demise of a single species, and in some instances, mass species extinctions, appear to be a consequence of genetically guided biological activity with "genes acting on the environment and the changing environment acting on gene selection;" interactions which result in cell and species death and the loss vs gain of genes which corresponds with the extinction of previous species and the evolution of new species which emerge from the old.

Thus extinction in many instances may be related to the same genetic mechanisms involved in embryogenesis and metamorphosis, where billions of cells die and others take their place, and where species shed one body, which dies, and replace it with another (Joseph 2000a, 2009d,e). Therefore, just as cells undergo programmed cell death during embryogenesis and metamorphosis, enabling, for example, a tadpole to shed its fish-like body to become a frog, or a crawling insect to destroy most of its body and then undergo a dramatic transformation to become a flying insect, that many species serve as a genetic bridge to a subsequent species and die out after having served a biological purpose such as by altering the environment through the secretion of various substances or by becoming a food source.

Therefore extinction is related to metamorphosis, embryogenesis, and is sometimes a form of evolutionary apoptosis and is under genetic-environmental regulatory control. "As a form of evolutionary apoptosis, extinction is in part a direct consequence of the same cellular mechanisms which lead to cell death; albeit at the level of an entire species" (Joseph 2009d)

Metamorphosis and Evolutionary Apoptosis

Birth, death, evolution and extinction, are part of the cycle of life. Like programmed cell death, extinction may be intrinsic to and necessary for the development, evolution, and metamorphosis of increasingly complex species (Joseph 2009e,f). Extinction, in some instances, may be considered a form of evolutionary sculpting and "apoptosis"--from a Greek word referring to leaves "falling off" a tree--and promotes the spread and diversification of new life.

Broadly considered, extinction may be due to a number of factors (Bradshaw and Brook 2009; Elewa & Joseph 2009; Raup 1991), including: **1)** "accidental causes" such as gamma rays, bolide impact and volcanic activity, whereby numerous species are wiped out by random uncontrolled catastrophic forces (Arens & West 2008; Elewa 2009; Jablonski 1994; Melott et al. 2004; Thompson and Crutzen 1988), or **2)** "biological causes" (Casadaveli 2005; Jones 2009; Poinar & Poinar 2008; Miller et al. 2004), including interactive biological (Ward 2009) environmental feedback mechanisms (Lovelock 2006) and genetic preprogramming leading to species death and eradication (Joseph 2009d,e). In the latter instances, extinction is linked to metamorphosis and evolutionary-apoptosis, with new species emerging from the old.

As a form of evolutionary apoptosis, extinction can be and is often tightly regulated at the genetic and cellular level (Joseph, 2009e), and often occurs in response to specific environmental (Lovelock 2006) and biological triggers (Ward 2009), all of which can and will initiate the mass death and elimination of specific species. Multicellular organisms which served as a genetic bridge to subsequent species, and which have fulfilled their biological purpose and provide no additional biological/environmental function, are destroyed by biologically/genetically regulated processes.

Biological Suicide

Over the course of the history of this planet billions of species may have been genetically programmed to commit biological suicide (Ward 2009); a consequence of the fact that genes act on the environment and the changing environment acts on gene selection and induces gene gain and gene loss and the evolution vs the extinction of species (Joseph 2000a, 2009a). The interaction between the environment and genetic activity, regulates the emergence of new species and the elimination of yet others--a form of evolutionary-apoptosis.

It is this genetically preprogrammed apoptosis, and the biological activity of viruses and prokaryotes, which may help explain why numerous distinct species evolved or became extinct during and after the first two global ice ages (Joseph 2009d,e) and during the Ediacaran age, only to die out (Amthor et al. 2003; Narbonne 2005), and then again during the Cambrian explosion only to become extinct (Palmer 1998; Westrop & Ludvigsen 1987; Zhuravlev & Wood 1996).

Likewise, if we accept the view of gradualists (e.g., MacLeod et al. 1997) vs the catastrophists (e.g., Alvarez 2008), preprogrammed apoptosis could explain why dinosaurs were progressively and selectively wiped out, such that by 65 million years ago, after a rein of over 200 million years, these species finally disappeared following bolide impact (Alvarez 2008; Arens & West 2008; Elewa 2009; Poinar & Poinar 2008) whereas others, such as mammals, recovered, continued to thrive, diversified, and became increasingly intelligent.

Apoptosis, Mitochondria, and Species Mass Death

Species may become extinct due to a variety of causes (Bradshaw & Brook

2009; Firestone 2009; MacLeod et al. 1997; Miller et al. 2004; Poinar & Poinar 2008) many of which are only tangentially related to the forces associated with natural selection. For example, species-destroying-diseases induced by bacteria, fungi, and viruses (Casadevall 2005; Devaraj 2000; Emiliani 1993; Gong et al. 2008; Poinar & Poinar 2008), may have little to do with natural selection, and everything to do with evolutionary apoptosis, and alterations in the host-genome or the genome of the pathogen, which enable pathogens to selectively target and kill off a specific host long after it has evolved (Flint et al. 2009; Norkin 2009).

Not uncommonly, diseases which have extinction-potential are transmitted vertically, from genome to genome via mitochondria (Engelstädter & Hurst 2007). Male-killing bacteria, for example, can hitchhike over thousands of generations, from mother to offspring, embedded within the mitochondrial genome (Jiggins 2003) and may begin destroying a species only after a related species has evolved. In the case of retroviruses they are transmitted within the nuclear genome, and will selectively sicken or kill specific species in response to as yet unknown biological and environmental triggers (Joseph 2009e). Genes which encode for retroviruses can be passed down for millions of years, embedded in the eukaryotic nuclear genome, and when expressed not only promote speciation and the evolution of new species but simultaneously eradicate others.

Interactions between the genome of mitochondria and the Eukaryotic nucleus, which are derived from microbes, and the biological activity of microbes and their effects on the environment, play a central role in apoptosis and the genetically programmed extinction of species (Joseph 2009e). In fact, stasis vs alterations within the mitochondrial genome have been directly linked to species extinction (Ballard & Rand, 2005; Gilbert 2008; Hofreiter et al. 2007), and mitochondria are directly implicated in apoptosis and cell death (Yin & Dong 2009).

Achae, bacteria, and Cyanobacteria all contributed genes to the Eukaryotic genome including the creation of the nucleus and the various compartments of the Eukaryotic cell. These internalized microbes act as "gatekeepers" determining which genes from other Prokaryotes and viruses may be inserted into the genome and which may be denied entry. Likewise, mitochondria appears to be a stripped down internalized microbe which underwent metamorphosis following significant increases in atmospheric oxygen, which had been excreted by Cyanobacteria. In consequence, some species of Eukaryote evolved and become more complex whereas yet others became extinct.

Mitochondria, therefore, not only promote evolution, but they can promote extinction (Ballard & Rand, 2005; Gilbert 2008; Hofreiter et al. 2007) by killing individual cells and the entire organism. And the same may be true of those internalized archae and bacteria which formed compartments and the nucleus; that is, they can promote evolution by allowing genes entry, or they promote death by denying entry to specific genes or by allowing bad genes and contagion producing viruses to infect the cell (Joseph 2009e).

As a form of evolutionary apoptosis, extinction is in part a direct consequence of the same cellular mechanisms which lead to cell death; albeit at the level of an entire species.

Apoptosis: Competition for Survival and Programmed Death

Competition for survival occurs not just at among species and individuals, but between cells. Consider, for example, the developing brain where billions of neurons proliferate and compete for functional representation with the losers dying out (Casagrande & Joseph, 1978, 1980; Joseph, 1999, 2000b, 2012; Joseph & Casagrande, 1978, 1980). This cellular competition leading to the death of millions of healthy neurons is due, in part, to the effects of the environment on gene activity, and enables the brain to differentiate, to become "fine tuned," and to function at optimal capacity. Alterations in and the changing environment turns certain genes on and others off and activates silent genes thereby promoting brain cell development, cellular growth, and the interactions and establishment of neural connections between brain cells resulting in the acquisition of new sensory perceptual functions and intellectual abilities (Casagrande & Joseph, 1978, 1980; Joseph, 1977, 1999, 2000b, 2012; Joseph & Casagrande, 1978, 1980; Joseph & Gallagher, 1980). Cells grow larger, establish new and more elaborate connections, whereas conversely, other cells atrophy and undergo cell death as a direct consequences of the environment.

Likewise, during human fetal development, billions of healthy cells compete, and losers continually die (Joseph 2000b). Cell death is a normal function of fetal development, effecting not just the brain, but, for example, fetal cells which form webs of thin tissue linking the fingers and the toes (Alberts et al. 2007). If not for cell death these appendages would be fused and webbed thereby leading to disability and reduced functional capacity. Instead of a webbed hand which would be appropriate if living in the sea, individual fingers separate as the tissues between them undergo programmed cell death such that a webbed appendage becomes a five-fingered hand.

Tissue and organ development are typically preceded by cellular proliferation, differentiation, and competition, with excess or unnecessary cells "pruned" away by apoptosis (Alberts et al. 2007; Yin & Dong 2009). It has been estimated that 50 billion cells die each day due to apoptosis in the average human adult (Alberts et al. 2007; Yin & Dong 2009). Thus, the brain is pruned of competitors and excess cells and the same is true of the body. However, cell death is necessary for the body to grow and to acquire new functions, and the same can be said of extinction events and those processes which lead to the emergence of new species. That is, species must die pruned from the tree of life which allows other species to develop and grow; extinction is a form of cell death but at the level of entire species (Joseph 2009e).

Evolution is a form of embryogenesis. Therefor like populations of cels, specific species are selectively targeted for eradication as if they are being "pruned" from the tree of life, thereby making way for the development and evolution of new species which take their place. Extinction is largely under genetic control. As to catastrophic events mass extinctions account for less than 5% of all species who have disappeared from the face of the Earth (Erwin 2001). In fact, in instances where many species are driven to near extinction, some recover whereas others die out (Erwin 1998).

Although extinction is often considered a byproduct of Darwinian natural selection and "survival of the fit," Darwin's hypothesis (Darwin 1859, 1871) can be best summed up as "survivors survive because they are fit, and because they survive this is proof that they are fit" --a view which is little more than circular reasoning based on hindsight. Darwin's tautology cannot be used to make predictions, cannot be scientifically tested, and can only be applied after-the-fact. It is impossible to predict as based on Darwin's theory, which species will evolve, which will become extinct or which species is fit (other than the fact that they exist). According to Prothero (1998) up to 50 billion species have become extinct over the life time of this planet, leaving less than fifty million alive today. This means over 99.9% of all species have become extinct! By Darwin's definition, then almost no species are fit.

Embryogenesis and Apoptosis

At the cellular level, what has been called "evolution" and "extinction" in some respects parallels embryonic development (Joseph 2000a,b, 2009d,e). For example, the common ancestors of humans and chimpanzees likely had a tail, as is the case with monkeys and many other species of primate-mammal. Human and chimpanzee embryos develop a tail which then undergoes cell death, atrophies and recedes with the remainder undergoing a secondary metamorphosis and becoming the coccyx (Keith 2009; Laubichler & Maienschein 2007; Robert 2008). Whales evolved from land mammals, but upon returning to the sea lost their hair and legs. Yet during embryonic development, whales become hairy and develop leg extremities which then undergo massive programmed cell death and disappear (Laubichler & Maienschein 2007).

Metamorphosis and Apoptosis

Be it fish, reptile, or human primate, the embryos of each of these species develop "gill slits" and gill pouches which in humans undergoes a major metamorphosis and becomes the pharyngeal arches and clefts of the throat and the canals of the middle ear (Laubichler & Maienschein 2007; Robert 2008). Yet a third pouch arising from the embryonic "gill slits" becomes the parathyroid and thymus glands (Keith 2009; Robert 2008) which are involved in the metabolism of calcium. Calcium plays a significant role in apoptosis (Mattson & Chan 2003) and plays a significant role in the death and growth of cells, and the evolution and extinction of entire species.

Ontogeny does not replicate phylogeny. Nevertheless, many aspects of embryology and metamorphosis parallel physical events associated with extinction and evolution, and involve apoptosis and the death of billions of cells.

Embryological development (Carlson, 2013) can also be likened to metamorphosis, as both involve the proliferation and death of healthy cells, and the selectively turning on and off of specific genes and nucleotide sequences; and the same can be said of evolution (Joseph 2000a,b, 2009d,e). Metamorphosis typically involves mass cell death and a complete transformation with one body being replaced by a completely different body which at first may feed upon the dead and still living cells from its old body (Dufour, et al., 2012; Evans & Claborne 2005; Gilbert 2009; Heifman et al. 2009; Tackett & Tacket 2002)

For example, many species of insect first enter a larval followed by a pupa stage and then undergo transformations in morphology and physiology; growing wings, legs, reproductive organs and complex compound eyes which sprout from discrete enfolded pockets of tissue (Gilbert 2009). Depending on species, insect metamorphosis often involves holometabolism and complex gene-environmental interactions resulting in the death of many of the cells of the body. Holometabolism can be considered a form of cell-suicide. The insect's body will "digest" itself, with intact cells feeding upon the dead cells and then rapidly grow, undergoing a radical transformation into a seemingly completely different species.

A wide variety of species, including barnacles, bivalves, bryozoans, crabs, echinoderms, gastropods, mollusks, sea squirts, flat fish, flounder, eels, amphibians, and reptiles, undergo profound physical alterations after birth and infancy which are often accompanied by massive cell death, the metamorphosis of new cells and body parts and significant changes in behavior and habitat (Duellman & Trueb 1994; Dufour, et al., 2012; Evans & Claborne 2005; Gilbert 2009; Heifman et al. 2009; Tackett & Tacket 2002). And those dead cells and discarded body parts are often utilized as food sources for the new body which emerges.

Metamorphosis, therefore, results in profound and rapid alterations in gene expression and body structure as cells differentially die and are replaced, such that a completely different and more complex and advanced life form emerges from a body which essentially dies. The same can be said of embryogenesis and evolution. However, if metamorphosis took a million years, instead of a single season, the Darwinians would claim this as evidence of random mutation and natural selection, when in fact the entire process is due to genetic-environmental interactions and is under precise, genetic regulatory control.

Genetic Apoptosis

When a new species emerges, it is not completely unique but shares genetic, skeletal, neurological, and other physical similarities with other species, including ancestral species who have been eradicated. These commonalities include the

103

skeletal system, the eye, heart, body and brain, and the genes which code for these organs and structures (Callaerts et al. 1997; Gehring & Ikeo 1999; Hadrys et al. 2005; Quiring et al. 1994; Salvini-Plawen & Mayr 1977; Sodergren et al. 2007). There are genetic commonalities and the presence of hundreds and thousands of highly conserved genes which in modern species, including humans, are shared with the genomes of even distantly related species (Snel et al. 2002; Mirkin et al. 2003; Kunin & Ouzounis 2003; Koonin 2003; Mushegian 2008; Bejerano et al. 2004), and which can be traced to common ancestors which died out over a billion years ago (e.g., Hedges & Kumar, 1999; Joseph 2009a; Wang et al. 1999).

Many of these highly conserved genes were were acquired from archae and bacteria (Yutin et al. 2008; Esser et al. 2004, 2007; Rivera & Lake 2004) via horizontal gene transfer, possibly over 4 billion years ago (Joseph 2009a).

Individual genes also undergo apoptosis often following an episode of gene proliferation and the duplication of the entire genome (Aravind et al. 2000; Dehal & Boore 2005; Durand 2003; Katinka et al. 2001; Moran 2002; Scannell et al. 2007; Wolfe & Shields 1997). Genes proliferate and compete for expression, with losers dying out and undergoing genetically programmed apoptosis (Joseph 2009a).

The entire genome has been duplicated repeatedly over the course of evolution, growing in size with genes proliferating and others dying out. Whole gene and whole genome duplication, coupled with gene loss and gene deletion, date back to the emergence of the first eukaryotic cells or their ancestors (Makarova et al. 2005).

Gene proliferation coupled with gene loss is a major feature of evolutionary processes which have given rise to distinct species and lineages (Aravind et al. 2000; Moran 2002). Genome analysis has revealed the extensive loss of genes after whole genome duplication in chordates (Dehal & Boore 2005; Durand 2003; McLysaght et al. 2002), plants (Soltis et al. 2008; Tuskan et al. 2006), and yeasts (Katinka et al. 2001; Scannell et al. 2007; Wolfe and Shields 1997).

Gene loss is a common phenomenon and appears to play an important role in shaping genome content (Snel et al. 2002) and the evolution of new species (Joseph 2009b,c). The extent of gene loss can be dramatic, and it can occur relatively rapidly under a strong selective pressure (Baumann et al. 1995) such that, not only do new species acquire additional genes, but they lose yet others which had belonged to the genomes of ancestral species which became extinct. Substantial gene loss has occurred in all phylogenetic lineages (Snel et al. 2002; Mirkin et al. 2003); indicating that gene eradication plays a significant role not just in evolution, but extinction.

The eradication of the original gene may also play a role in the expression of the duplicate which may jump to a new position within the genome, thereby altering the genetic code (Joseph 2009b,e). Some of these duplicate genes appeared to have been freed from inhibitory restraint and were able to undergo an accelerated rate of sequence change thereby inducing the rapid evolution of new characteristics and abilities (Seoighe et al. 2003). After genome duplication followed by gene deletion, the duplicate or original genes, now freed of suppressive constraints, can express an already encoded function ("neofunctionalization") which had been repressed (Conant & Wolf 2008).

New species do not evolve new genes. Genes are inherited, and these genes may be duplicated, change their position in the genome, exons and introns may be shuffled, and nucleotide sequences may be lengthened or shortened, such that genes inherited from ancestral species may come to be silenced or remain repressed (Nichoh et al. 2008) until activated by changes in the environment or the genome (Joseph 2000a, 2009b,c,e). Likewise, following gene deletion, genes are freed of competitive restraint, and come to be expressed, whereas others may be lost or inhibited; thus new species emerge and others become extinct.

It is not uncommon for the new gene paralogs to retain or express distinct subsets of the original functions of the ancestral gene whereas the rest of the functions differentially deteriorate (Lynch & Force 2000; Lynch & Katju 2004). However, when ancestral genes are pruned from the genome, the cells, tissues, organs, and structures they code for also become modified, or are lost and undergo apoptosis, paralleling the genetic mechanisms regulating embryogenesis and metamorphosis.

Therefore, the shaping of a new species involves not just the emergence of new traits, but the activation or suppression of specific genes that code for specific physical features, coupled with gene gain and gene loss (Joseph 2009a). The result is the modification or elimination of certain physical characteristics associated with an ancestral species which may become extinct. Therefore, cells and genes that can give rise to specific physical characteristics in a progenitor species may be lost in a subsequent species, and this is due, in part, to apoptosis and selective cell and gene death. Due to the selective preservation or loss of specific genes and cells, new species can be rapidly built from the genome of a progenitor species which ceases to exist. Hence, there is no need for an intermediary species to link the old with the new. Evolutionary change, like metamorphosis, can be rapid and dramatic. New species, therefore, emerge from ancestral species in genetically preprogrammed quantum leaps, whereas ancestral species are pruned from the tree of life.

Apoptosis and alterations within species-specific genetic codes, are an integral aspect of extinction and evolution.

The extinction of species, like programmed cell death and the events associated with metamorphosis and embryological development serve a genetically programmed biological purpose. If this were not so, a tadpole could not become a frog, a butterfly would remain a caterpillar, and an embryo could not become an infant, a child could not become an adult, and the ability of new and increasingly intelligent species to emerge, separate, diversify, and invade empty niches, would be significantly hindered.

Genetic Seeds: Archae, Bacteria, Eukaryotic Evolution, Mitochondria

At the level of DNA, all species are linked and can be traced backward in time to common ancestors (Bejerano et al. 2004; Gu 1998; Koonin 2003; Koonin et al. 2004; Koonin & Wolf, 2008; Mirkin et al. 2003; Mushegian 2008; Nei et al. 2001; Peterson et al. 2004; Snel et al. 2002; Wang et al. 1999). Therefore, all species, and their DNA, can ultimately be traced to the first creatures to arrive on Earth (Joseph 2000a, 2009a,b). However, what this implies is that the DNA of the first Earthlings also leads backwards in time, to DNA-based life forms which lived on other planets (Joseph 2000a, 20009a,b).

Therefore, based on a genomic analysis, and from the perspective of meta-morphosis and embryology, rather than a random evolution, evolution may be a form of metamorphosis, the replication of creatures that long ago lived on other worlds (Joseph 2000a, 2009a). That is, just as an apple seeds contains the genetic machinery for growing an apple tree if planted in the right environment, the first (and subsequent) microbes to arrive on Earth contained the genetic machinery for growing the tree of life, a forest of life, and for altering the womb of the planet so this "tree" and this forest could grow branches and bear fruit in the form of increasingly complex species. And just as most trees lose branches, twigs, leaves, and fruit before they are ripen, thereby providing other branches and leaves with greater access to nutrients and sunlight, species also fall from the tree of life and become extinct; and their extinction is to the benefit of other species, all of whom, be they those who survive and those who die out, share many of the same genes which can be traced to Prokaryotes and viruses.

At the level of DNA, all species are linked and can be traced backward in time to common ancestors (Bejerano et al. 2004; Gu 1998; Koonin 2003; Koonin et al. 2004; Koonin & Wolf, 2008; Mirkin et al. 2003; Mushegian 2008; Nei et al. 2001; Peterson et al. 2004; Snel et al. 2002; Wang et al. 1999). Therefore, all spe-cies, and their DNA, could be considered a supra-organism, linked by DNA. And just as any and all organisms, beginning as embryos or tiny roots emerging from various seeds, undergo programmed cell death, various species are also pruned from the tress of life.

Extinction and "evolution," like embryogenesis and metamorphosis, ae under genetic control, and these genetic mechanism were encoded into the genomes of the first creatures to arrive on this planet; microbial life forms which began biologically engineering the environment, which acted on genes inserted into the Eukaryotic genome. Considered as a supra-organism, all genes and all species are linked, and therefore, collectively, this supra-organisms as a whole is greater than the sum of its parts, many of which must be shed so that the supra-organism can continue to develop, grow, and continue onward to the next stage of collective metamorphosis.

Genetic Seeds of Destruction: Evolution & Extinction

Life gives birth to life, and stars give birth to stars in an endless cycle of death and rebirth. It is a cosmic dance which may have been ongoing for all eternity. It is the death of a star, and the mass extinction of the species which populated its planets, which may have given birth to our world and life on Earth (Joseph 2009b). Life on Earth came from other planets.

Every creature alive today, and their DNA, can be traced backward in time to the first viruses and prokaryotic life forms to take root on Earth (Joseph 2009a,b; Woese 1994, 2002); i.e. archae, bacteria and blue-green algae--also known as Cyanobacteria. And the these first denizens of Earth, or their immediate ancestors were delivered via comets, asteroids, meteors, and planetary debris. And these first, and subsequent arrivals, these sojourners from the stars contained within their genomes what could be likened to "genetic seeds"--the genetic potential for the evolution of all subsequent species, leading to woman and man (Joseph 2000a; 2009a); and not just their evolution, but their extinction. All cells and all multi-cellular species contain the "genetic seeds" of their own destruction.

These first Earthlings (archae, bacteria, viruses) contained the genes and genetic information for altering the environment, the "evolution" of multicellular eukaryotes, and the metamorphosis of all subsequent species and their extinction. Some of the genes which were transferred to the Eukaryotic genome by Prokaryotes and viruses, included exons, introns, transposable elements, informational and operational genes, RNA, ribosomes, mitochondria, and the core genetic machinery for translating, expressing, and repeatedly duplicating genes and the entire genome (Charlebois & Doolittle 2004; Dehal & Boore 2005; Harris et al. 2003; Koonin et al. 2004; Lynch et al. 2001; McLysaght et al. 2002). Prokaryotes (archae and bacteria) and viruses provided Eukaryotes with the regulatory elements which control gene expression and which have repeatedly duplicated individual genes and the entire genome thereby enabling the Eukaryote gene pool to grow in size. However, among these genetic mechanisms, were the tools for eliminating genes and entire species (Joseph 2009e).

After Eukaryotes began to evolve, viral and Prokaryotic genes continued to be transferred to the the Eukaryotic genome, and these many of the donated genes were "silent" and were transmitted via the germ line to subsequent generations and to subsequent species, and coming to be expressed in response to biologically engineered environmental influences and specific genetic regulatory signals, often in busts of explosive evolutionary change, as typified by the Cambrian Explosion (Joseph 2009d).

DNA-equipped cells biologically alter the environment and liberate, secrete and manufacture various products, e.g. methane, oxygen, calcium carbonate, sulphate, ferrous iron, etc., all which act on gene expression, generating for example, eyes, hearts, bones, bodies, and brains (Joseph 2009d). These genes coding for brains, bones and eyes did not evolve. They were inherited from boneless, brainless, heartless ancestors, and then activated or suppressed by the changing environment including the buildup of atmospheric oxygen and massive amounts of calcium which leached into the sea (Joseph 2009d); oxygen and calcium being produced largely by Cyanobacteria and Cyanobacteria genes contributed to Eukaryotes.

However, calcium (Mattson & Chan 2003) and oxygen (Yin & Dong 2009) not only promoted the metamorphosis of new life forms, but trigger cell death and possibly the biological suicide of entire species.

Calcium, Cyanobacteria & Exinction

Cyanobacteria have made profound contributions to the evolution of the external biosphere and to the internal biosphere and genome of the Eukaryotic cell. Not just oxygen and calcium, but the intra-cellular endoplasmic reticulum, owe their origins largely to Cyanobacteria. The endoplasmic reticulum also processes the calcium secreted by Cyanobacteria, and it directly interacts with mitochondria to store

Ca^{2+} store thereby enabling organelles to rapidly and effectively respond to cellular Ca^{2+} signals and which stimulates mitochondrial Ca^{2+} uptake machinery (Filadi et al., 2012) which in turn physically affects the various intracellular compartments and membranes. These intracellular networks which control Ca^{2+} dynamics not only contribute to the survival and continuing evolutionary metamorphosis of species, but to pathological conditions which result in their extinction.

The first Snow Ball Earth was orchestrated biologically via the activities of Cyanobacteria and methanotrophs whose secretions of oxygen and consumption of methane resulted in global freezing. Calcium and oxygen levels fluctuated and the changing environment then acted on gene selection, triggering the next stage of metamorphosis as well as ushering in waves of mass extinction; "pruning" of the tree of life. And then temperatures began to rise, due again to biological activity, and yet another wave of mass extinction coupled with the evolution of new species ensued. However, those who died became sources of food, thus providing for those who were yet to evolve.

Following the close of the Marinoan and then the Gaskiers glaciation and the warming of the planet, an explosion of life ensued (Condon et al. 2005; Peterson & Butterfield 2005) including the evolution of megascopic Ediacarans (Narbonne 2005; Narbonne & Gehling 2003). However, with the oceans flooding in calcium, and by 540 mya, and the onset of the Cambrian Era, the Ediacaran age would come to a close and the Ediacaran would become extinct.

The Ediacaran were killed off by a number of factors, and which may have included the rise in oxygen and calcium levels, both of which acted directly on mitochondria, the endoplasmic reticulum and gene selection. Cells absorb and secrete Ca^{2+} and calcium receptors and sensors are located throughout the body. However, increased intracellular CA can also trigger apoptosis (Mattson & Chan

2003) and thus cell death, and possibly, contributed to the eradication of entire species (Joseph 2009e). Thus, as metazoans began to proliferate, yet other species suffered a progressive and massive die off, including the Ediacaran fauna which became extinct, bringing the Ediacaran era to a close.

The endoplasmic reticulum also processes the calcium secreted by Cyanobacteria, and then synthesizes collagen which stimulates calcium binding. Calcium plays a central role in cellular proliferation and differentiation, cell to cell adhesion and fusion, apoptosis, and programmed cell death (Brown and MacLeod 2001; Cheng et al. 2007). In the absence of Ca, or in response to excess levels of intracellular Ca, cells stop aggregating, embryos fail to adhere, cell aggregates disintegrate, and bones become soft and easily break. Thus fluctuations in calcium levels can trigger mass extinction beginning with the mass die off of embryos.

The intracellular structures which can determine and control intracellular calcium levels are the mitochondria-endoplasmic reticulum network which in turn can effect the physical integrity of the cell and its various compartments (Filadi et al., 2012). The mitochondria and the cells nucleus and various compartments are comprised of stripped down internalized Prokaryotes which transferred genes to the Eukaryote genome. Therefore, complex genetic interactions involving Cyanobacteria, genes contributed to the Eukaryotic genome by Prokaryotes, and structures within the cell can determine if the cell dies, if embryos fail to form, if bones become fragile, and if entire species become extinct. Stasis vs alterations within the mitochondrial genome have been directly linked to species extinction (Ballard & Rand, 2005; Gilbert 2008; Hofreiter et al. 2007), and mitochondria are directly implicated in apoptosis and cell death (Yin & Dong 2009).

The Ediacaran extinction, the Marinoan and Gaskiers extinction, the Sturtian extinction, and the Paleoproterozoic extinction, all of which were related to Cyanobacteria activity and fluctuations in oxygen and calcium levels, may therefore be examples of biologically induced and genetically controlled evolutionary apoptosis (Joseph 2009e). These species were "shed from the tree of life" after having served an an evolutionary bridge to subsequent species including those which flourished during the Cambrian Explosion (Joseph 2000a).

Over the course of this planet's history, innumerable microbes and millions of species have genetically engineered the climate, triggering episodes of global freezing and global warming and flooding the oceans and atmosphere with a variety of gasses and secretions, thereby selectively destroying billions of life forms in favor of their more advanced cousins who were perfectly adapted for a genetically altered world which had been biologically prepared for them (Joseph 2009e). Genes act on the environment which acts on gene selection, giving rise to increasingly complex species, while simultaneously driving others to biological extinction.

All cells and all multi-cellular species contain the "genetic seeds" for their evolution and their extinction. And among these "genetic seeds" of life and death are mitochondria.

Mitochondria, Oxygenation, Glaciation

Mitochondria, as a distinct entity within eukaryotic cells, did not arise until between 2.3 to 1.8 bya (Mente & Martin 2008) when oxygen had begun to enrich the atmosphere (Barleya et al. 2005). It was increased levels of oxygen which impacted the stripped down-bacterial-proto-mitochondrial invader and Eukaryotic genome, resulting the metamorphosis of mitochondria.

It was during this time that the Earth became glaciated, fueled by oxygenic photosynthesis by Cyanobacteria (Evans et al. 1997; Kirschvink, et al. 2000). This rise in oxygen has been referred to as the Paleoproterozoic "Great Oxidation Event" (~2.3 to 2.0 Ga), when atmospheric oxygen may have risen to >1% of modern levels, and which coincided with the onset of the Proterozoic, and the first "snow ball Earth." Thus not just the metamorphosis of mitochondria but the Earth's earliest ice age are linked to the rise of oxygen in Earth's atmosphere; a time period when many species became extinct and others evolved (Joseph 2009e; Raup 1991).

Prior to the first global ice age, triggered by prokaryotic biological activity, the Earth had been experiencing a period of global warming from a greenhouse effect, an organic haze, created by increasing levels of H_2 and CH_4 secreted by methane-producing microbes as a waste product (Brocks et al. 1999, 2005). The high levels of methane also acted on gene selection, and archae, known as methanotrophs and methylotrophs, began to proliferate. These were methane eaters, and in ever growing numbers they metabolized and broke down methane, as demonstrated by the presence of hopanes and high relative concentrations of 2α-methylhopanes in Archean rocks (Brocks et al. 2005). These methane eaters had also contributed genes to Eukaryotes which were also utilizing alternative energy sources.

As methanotrophs proliferated, methane levels were reduced, thus dissipating the organic haze and allowing more sunlight to strike Earth; and this resulted in increased photosynthesis by Cyanobacteria and other microbes. By 2.45 bya, oxygenic photosynthesis had become widespread (Brock et al. 2005; Buick 2008) and atmospheric oxygen levels continued to rise (Bau et al. 1999; Kirschvink et al. 2000, 2008) to values between 0.02 and 0.04atm (Holland 2006).

Oxygen also breaks down methane. The presence of even small amounts of O_2 in the atmosphere would have been associated with decrease in CH_4 and this decrease would have caused the planet to rapidly cool (Kasting and Ono 2006; Young et al. 1998).

Increased oxygen levels beginning around 2.4 bya, eventually shut down sulphur MIF production and caused a rapid and drastic decrease in atmospheric CH_4, thus triggering glaciation. That is, increased levels of O_2 --which were secreted as a waste product by Cyanobacteria and Eukaryotes with Cyanobacteria genes-- acted to oxidize sulphide, such that dissolved sulphate levels increased just as

O_2 levels increased. Both began to build up in shallow marine sediments which resulted in decreases in methanogenesis and reductions in the CH_4 and caused significant reductions in atmospheric methane (Kharecha et al. 2005; Pavlov et al. 2000, 2001, 2003). Temperatures began to drop. The increased levels of sulphate in turn triggered a proliferation of sulfur-eating bacteria, which caused a drawdown in H_2 and CH_4, a consequence of bacterial sulphate reduction (Kasting & Ono 2006).

Photosynthetic bacteria were also employing H_2, H_2S and/or Fe^{2+} to reduce CO_2 to organic matter (Pierson 1994). Reductions in methane coupled with reductions in CO_2 and increases in oxygen secondary to photosynthesis, eliminated the greenhouse effect and accelerated the cooling of the planet which began to freeze (Roscoe 1969, 1973), creating the first "snowball Earth" referred to as the "Makganyene" glaciation. However, Photosynthetic bacteria which were employing H_2, began to die as did those organisms which were not adapted to freezing temperatures.

Thus, biological activity of methane-eating and photosynthesizing microbes and the release of oxygen into the atmosphere triggered the first global ice age, such that by 2.2 bya much of the Earth and its oceans were frozen or covered with ice and snow (Evans et al. 1997; Kasting and Howard, 2006; Kirschvink, et al. 2000; Roscoe 1969, 1973), creating the first "snowball Earth" and the first mass extinction. The first mass extinction was orchestrated by Cyanobacteria and those photosynthesizing microbes with Cyanobacteria genes.

As the Earth continued to freeze, these blankets of snow and layers of ice also provided protection against UV rays, but allowed light penetration (McKay 2000). This enabled photosynthesizing Cyanobacteria to proliferate near the surface (Cockell et al. 2002; Cockell and Cordoba-Jabonero 2004). Cyanobacteria secreted even more oxygen into the atmosphere, thus maintaining the low temperatures.

The changing environment acts on gene activation and inhibition, and increased oxygen levels triggered the metamorphosis of mitochondria (Joseph 2009d).

The genomes of all extant eukaryotes contain genes which can be traced to ancestors that possessed the α-proteo-mitochondrial bacterial endosymbiont that gave rise to the mitochondria (Embley 2006; van der Giezen & Tovar 2005). Presumably, the genes of this symbiont, and those genes donated by archae and Cyanobacteria, underwent metamorphosis in response to the increasing levels of oxygen in the atmosphere, becoming a mitochondria and creating additional cellular compartments.

Therefore, we see that species of Prokaryotes which transferred genes to Eukaryotes, acted on the environment, which acted on the genes donated by prokaryotes, thereby triggering the metamorphosis of mitochondria and multicellular eukaryotes due to the oxygenation of the planet. However, innumerable microscopic species, both Eukaryotic and Prokaryotic, were also driven to extinction by these biologically engineered changes in the environment which created alternating periods of global warming and global freezing (Joseph 2009c,d,f).

Environmental stresses, including extremes in temperature and fluctuations in oxygen , directly act on mitochondria (Alberts et al. 2007; Ballard & Rand 2005; Yin & Dong 2009; Zhuravlev & Wood 1996) and can promote or inhibit gene expression (Nagata 2000; Rutherford & Lindquist 1998; Waterland and Jirtle, 2003; Wolff et al. 1998) and thus induce evolutionary change (Rutherford & Lindquist, 1998), and extinction. That is, changes in the environment can promote cellular proliferation, or apoptosis and cell death (Alberts et al. 2007; Yin & Dong 2009) and the extinction of entire species.

Death can also serve life. The first microbial mass extinction resulted in the formation of thick layers of carbohydrate enriched organic matter, thereby providing nutrients for yet other species who were yet to evolve. Dead microbes also provided fodder for surviving microbes who would significantly impact the environment via methanogenesis and the degradation of organic matter (Holland 2006; Joseph 2009d).

Methanogens and multicellular eukaryotes equipped with mitochondria, began to flourish and organisms with more than 2-3 cell types appeared (Hedges et al. 2004). This increase in energy availability (oxygen) and the ability to extract it (mitochondria) conferred major advantages for the eukaryotic host which became increasingly complex and expanded in size and which could now invade and colonize new environments. Oxygen breathing species proliferated, and others were shed from the tree of life.

Mitochondria and Evolutionary-Apoptosis

The mitochondria which changed the world and enabled Eukaryotes to evolve from simple cell to woman and man, also play a significant role in evolutionary-apoptosis and programmed species death. The genome of the mitochondria, in fact, contains the instructions for the manufacture of molecules which trigger cell suicide (Chiarugi & Moskowitz 2002; Yin & Dong 2009) thereby initiating the death of the organism, and possibly the eradication of an entire species.

For example, just as mitochondria reacted to increased oxygen and calcium levels, and the changing environment, by enabling new species to emerge and conquer the oxygenated environments which had been fashioned for them, mitochondria also react to environmental stresses, including temperature extremes, by manufacturing and releasing a cascade of chemicals, including caspase activators, which regulate and trigger the massive die off of cells (Brüne 2003; Chiarugi & Moskowitz 2002; Popov et al. 2002).

The death of an organism may be due injury, disease, profound environmental change, and the incapacity of cells to function; and central to cell death is the mitochondria and the chemicals manufactured by its genome which induce cell suicide. The biochemical cascade released by mitochondrial activity, can exert profound morphological changes impacting the cell membrane, causing cells to atrophy, fragmenting the cellular nucleus, and attacking, fracturing, and destroying DNA (Brüne 2003; Chiarugi & Moskowitz 2002; Nagata 2000; Kumar et al. 2009; Popov et al. 2002).

Consider the events leading up to and following the first snow ball Earth which involved not just temperature extremes but alterations in UV and other radiation. Changes in heat, oxygen, and radiation exposure also induce apoptosis (Kumar et al. 2009; Yin & Dong 2009). Hence, innumerable species were eliminated, others prevailed, and yet others emerged: a consequence of the changing environment acting on gene selection and cell function.

These events were triggered and regulated biologically and led to the emergence of new species and the extinction of others, who were removed from the tree of life.

Cells in fact have "death receptors" along their surface which trigger procaspase activation which can kill cells (Kumar et al. 2009; Yin & Dong 2009). Killer lymphocytes produce Fas ligand proteins which bind to death receptor proteins thereby activating procaspases and inducing apoptosis. In this way cells can kill themselves by triggering procaspase aggregation and activation from within the cell.

Stress, disease, or injury can also trigger the creation of procaspases which induce caspase cascades which cleave together and become activated, triggering and even amplifying a cell's death signal and spreading death throughout the cell and the organism (Kumar et al. 2009; Yin & Dong 2009).

However, these pathways also involve the mitochondria which releases cytochrome c into the cytosol, where it binds with and activates an adaptor protein called Apaf-1 which also triggers procaspase activation thereby accelerating and amplifying the caspase cascade and ushering in the mass suicide of cells (Kumar et al. 2009; Yin & Dong 2009).

Disease and viral organisms also trigger apoptopic cell death (Everett & McFadden 1999; Hay & Kannourakis 2002; Polster et al. 2004; Teodoro & Branton 1997) and thus the death of the organism or an entire species. In fact, viruses can also act on genetic regulatory mechanisms, such as p53, by inhibiting its ability to bind with proteins or engage in transcriptional activity (Wang et al. 1995). If the immune system, or master regulatory genes, such as p53, are targeted (Wang et al. 1995), not just an individual organism, but an entire species may be wiped out.

However, viruses are selective and only attack the genomes of specific tissues or specific hosts (Flint et al. 2009; Norkin 2009), thereby selectively wiping out the members of specific species while leaving others unscathed. They must also be allowed entry by Prokaryotic genetic gatekeepers, that is those stripped down Prokaryotic invaders which formed the nucleus and compartments within every Eukaryotic cell. It is also a Prokaryote which underwent metamorphosis and became a mitochondria. Thus, Prokaryotes have biologically altered the environment which acts on genes transferred into the Eukaryotic nucleus, as well as on mitochondria, the result of which is some species become extinct and others continue to evolve.

Viruses selectively target specific cells and specific species. The viral key must fit the genetic lock of the host and these viral invaders must be allowed entry by the internalized Prokaryotic gatekeepers. If the match is not perfect, the result can be death. However, the gatekeepers also allow viruses to invade thereby triggering mass death of a species.

If fact, viruses can simultaneously promote the evolution of one species, while targeting another for extinction. That is, viruses can also prevent apoptosis (Teodoro & Branton 1997), such that, whereas one species is impacted, another may be protected by the same viral pathogen (Flint et al. 2009; Norkin 2009).

Again, these complex interactions should not be viewed as agents of chance but from the perspective of and relative to a supra-genetic organism undergoing embryogenesis and metamorphosis. Once a species serves its genetic purpose, it may be shed from the tree of life which in turn continues to grow and bare evolutionary fruit.

Extinction, Evolution, and the Supra-DNA-Organism

At the level of DNA, all Earthly-life forms are linked a consequence of genes which have been continually horizontally transferred between Prokaryotes, Eukaryotes, viruses. Eukaryotic cells has been repeatedly invaded by Prokaryotes and viruses, with invading Prokaryotes acting as genetic gatekeeper by forming protective compartments and the nucleus and with other invading Prokaryotes becoming the mitochondria. All those who invade Eukaryotes also transfer genes to the Eukaryotic genome.

The exchange of genes is not random or due to chance, but under genetic regulatory control--much like the interactional precision of the inner workings of a digital clock. However, in this case, the "clock" is alive and can be likened to a supra-DNA-organism, or a giant all encompassing cellular entity whose component parts which include all species, interact according to biological laws and principles to maintain the health of the organism and to promote not just its survival, but its continual growth. And that growth may involve the shedding of "cells" and components parts which are no longer useful to the supra-organism. That is, not just evolution, but the extinction of species is under genetic control (Joseph 2009e).

Therefore, not only are modern day genes linked to ancestral species including the first and subsequent life forms to arrive on Earth, but intra- and interspecies DNA transfer, exchange, and interaction, raises the possibility that genetically all of Earthly life are components of a single, planet-wide, supra-DNA organism whose growth, evolution, and extinction, is coordinated and guided by genetic mechanisms in interaction with the biologically engineered environment.

Alien Viruses, Genetics, Cambrian Explosion: Humans

For the first 4 billion years most organisms were microscopic and Prokaryotes outnumbered all other forms of life, just as they do today (Staley et al. 1997). These Prokaryotes also donated essential genes to the Eukaryotic genome; and this exchange was under genetic regulatory control and served a biological purpose: evolution and metamorphosis. Subsequently, the activity of these Prokaryotes, and their genes, had a profound effect upon the growth and evolution of the tree of life; that is, on those genes donated to the Eukaryotic genome.

Just as a tree sheds leaves and branches as it grows, species are also shed in favor of continued evolutionary growth. Therefore, we see that photosynthesizing prokaryotes secreted massive amounts of oxygen and other chemicals into the environment; and the results were many including the metamorphosis of mitochondria and which led to the evolution of increasingly complex life forms, as well as the extinction of other species.

Thus, these genetic mechanisms led to the growth and evolution of what could be likened to a supra-DNA-organism which has orchestrated its own survival as well as the development of its various component parts, the evolution of new branches within the forest of life, the shedding of branches which have served their purpose and are no longer useful, and the metamorphosis of increasingly complex and intelligent creatures that long ago lived on other worlds.

7. Evolutionary Metamorphosis, Embryogenesis, and the DNA-Supra Organism

Genetic Seeds of Life: Viruses and Microbes

Viruses, microbial creatures and their DNA, are perfectly adapted for traveling from planet to planet and from solar system to solar system. Microbes and viruses act as interplanetary genetic messengers and through horizontal gene transfer (HGT) they acquired and their descendants inherited the genes which enable them to survive in almost any environment such as might be encountered on other worlds or in the wilds of space.

Be it an infinite cosmos or a Big Bang created universe, as microbes and viruses were cast from world to world, they transferred and exchanged genes with the denizens of these other planets including Eukaryotes and their fellow travelers. Genes were exchanged via HGT utilizing the same genetic mechanisms which are common among the microbes and viruses of Earth.

Various life forms may be ejected into space secondary to impacts by meteors, asteroids, and comets, or dramatic increases in the strength of solar winds. And as the descendants of these space-ferrying microbes and viruses fell upon other habitable planets much older than our own, they would have acquired, through HGT, genes coding for advanced and complex characteristics, features and traits, and those shaped by natural selection and the changing environments. Viruses began to store up vast libraries of genes and even multiple genomes.

As they were cast from planet to planet, solar system to solar system, and even to different galaxies, microbes and viruses would have contact with all manner of species, some of which, as is the case for Earth, had evolved into fish, amphibians, reptiles, dinosaurs, birds, repto-mammals, mammals, primates, and a variety of species similar to a wide variety of pre-humans such as *Australopithecus*, H. *habilis*, H. *erectus*, and archaic H. *Sapiens*. On habitable Earth-like planets they would have encountered species similarly if not almost identical to modern humans, and on worlds billions of years older than our own, they would have encountered intelligent life which had long ego evolved beyond those of the "modern" humans of Earth.

Nearly 4.6 billion years ago these microbes which were accompanied by viruses and their vast genetic libraries, arrived on this planet. And then these space-journeying microbes began to terraform the Earth, and to evolve.

Much of the human genome consists of genes inserted by viruses and Prokaryotes, including genes coding for the human brain, much of the remainder of the human genome can be attributed to regulatory activity of genes inserted into the Eukaryotic genome by these species beginning four billion years ago Thus, evolution leading to humans, has been guided by genes inserted by microbes and viruses whose own ancestry can be traced to yet other microbes and viruses which journeyed here from other planets. This indicates that similar genetic interactions leading to similar evolutionary progressions must have taken place on other worlds which are much older than Earth.

An analysis of the microbe, viral, and Eukaryotic genome and the evolutionary progression which has taken place on this biologically engineered planet leads to this conclusion: The first viruses and Prokaryotes to arrive on Earth contained the genetic instructions for the metamorphosis of all of life, and some of these genes were transferred to and gave rise to the multi-cellular Eukaryotic genome.

Just as an apple seed contains the genetic instructions for the development of an apple tree, these extraterrestrial genetic seeds of life contained the genetic instructions for the tree of life, a forest of trees, and for every creature which has walked, crawled, swam, or slithered across the Earth.

However, these "genetic seeds" also contained the genetic information and those genes which altered the environment, pumping, for example, oxygen and calcium into the biosphere; activities directly linked to Cyanobacteria which also contributed numerous genes to the Eukaryotic genome. Therefore, it could be said that Prokaryotes altered the environment which acted on those genes provided to the Eukaryotic genome by viruses and Prokaryotes, with viruses acting as a genetic storehouse which selectively targets specific species and tissues.

Thus, genes act on the environment, and the biologically altered environment acts on gene selection, thereby expressing traits which had been encoded into genes acquired from life on other planets.

Evolution, Metamorphosis, Embryogenesis

The implications are momentous: The first and subsequent microbes and viruses (and their genetic luggage) to fall to Earth engineered the activation and expression of those genes which they had inserted into varying hosts, resulting in a step-wise sometimes leaping progression, leading from simple cells, to multi-cellular organisms, then fish, frogs, reptiles, and finally humans.

These evolutionary events were not random acts of chance. Evolution is a form of metamorphosis and embryology. Indeed, metamorphosis, embryology, and development, are all under precise genetic control, why should "evolution" be any different?

Metamorphosis and embryogenesis are genetically regulated; guided and controlled by genetic-environmental interactions. The same is true of what is called "evolution" which, like embryogenesis, also began with DNA invasions into single cells and led to multicellularity. However, rather than 9 months to generate a complex organism beginning with a single cell and which leads to embryo, fetus, neonate... it takes billions of years to grow a human from a single cell.

What has been called "evolution" is under genetic regulatory control, in coordination with the biological activity of innumerable life forms which genetically engineer the environment which acts on the genomes of innumerable hosts. Genes act on genes, genes act on the environment, and the altered environment acts on gene selection, thereby giving rise to an evolutionary progression from simple cell to sentient intelligent being, each evolving into a world which has been genetically prepared for them.

The "genetic seeds of life" which swarm throughout the cosmos, contained the genetic instruction for the metamorphosis of all life on this planet. This does not mean that evolution and metamorphosis leading to modern humans was genetically pre-determined. Rather, the life forms that have evolved on Earth are just a sample of life's manifold evolutionary possibilities.

Although the generation of variations within a single species can be explained by Darwinian notions of "small steps", the generation of completely unique species on Earth took place by quantum leaps without intermediary forms. This was made possible by the inheritance of silent genes which coded for traits such as hearts, bones and brains; silent genes inherited from ancient heartless, boneless, brainless ancestral species whose genetic ancestry can be traced backwards in time to over 10 billion years, long before life appeared on this planet--life forms whose own ancestors hailed from other planets, other solar system, and perhaps other galaxies.

What has taken place on Earth represents not a random evolution, but the metamorphosis and replication of living creatures which long ago lived on other planets.

The Genetic Seeds of Life

The "genetic seeds of life" flow throughout the cosmos, and embedded in the genomes of these seeds are genes, genetic material, and genetic instructions for the metamorphosis of every possible life form, of which life on Earth is merely a sample. Once on Earth these seeds began to interact, to share genes, and to engage in a genetically regulated, highly choreographed dance of life which in many way is reminiscent of the interactions between a sperm and ovum which fuse after the sperm burrows within and deposits its genetic cargo resulting in the creation of a multicellular Eukaryotic genome from which emerges an embryo which becomes a neonate... but in terms of the evolutionary processes which has has taken place on this planet, germinates a forest of life.

The germination of these *genetic seeds* and the evolution of the life they contained were not determined by genetics alone, but required specific external conditions, the most important of which include a liveable range of temperatures, water, oxygen, calcium, and a protective ozone layer. Thus the Prokaryotes which had transferred their genes to Eukaryotes labored to biologically alter the planet to provide the oxygen and range of temperatures necessary to promote the germination, embryogenesis, and metamorphosis of life.

Evolution on Earth could be likened to metamorphosis and embryology. However, rather than 9 months, it takes billions of years to grow a human from a single cell; and an integral aspect of that evolutionary growth is apoptosis, also known as extinction. Embryogenesis is under genetic control. Metamorphosis is genetically regulated. All aspects of development are guided and controlled by genetic-environmental interactions. Why should evolution be any different?

Evolutionary Embryogenesis

The complex genetic interactions involving Prokaryotes, viruses, and Eukaryotes, can be likened to the generation and development of a complex supra-organism whose component parts consist of genetic chimeras and multiple genomic mosaics. Like the first Earthly Eukaryotes and paralleling the progression that ensues after the fertilization of an ovum by a sperm, the progression leading to humans began as a single cell. However, instead of a single "sperm", the "ovum" of this *supra-organism* is invaded by multiple packets of DNA provided by numerous species of archae, bacteria, and viruses.

For example, and considered in the broadest terms, mammalian embryogenesis begins with the fertilization of the ovum which becomes a single celled zygot, or embryo (Carlson 2013). What follows next corresponds to some of the evolutionary principles of punctuated equilibrium: The zygot then undergoes a very rapid metamorphosis and quantum leaps in cellular differentiation and becomes multicellular and compartmentalized internally and externally and covered with various membranes; all compartments and membranes serving a protective function (Carlson 2013); again paralleling the compartmentalization of the first Earthly multicellular Eukaryotes.

As detailed in this text, archae and bacteria, Cyanobacteria in particular, labored to biologically engineer the Earth and atmosphere, creating a biosphere which promoted evolutionary development; what could be likened to the "womb of the planet" (Joseph 2000a). Single celled Eukaryotes developed multiple cell types, and expanded from 2 to 10 cells, albeit remaining microscopic and not growing significantly in size until after they became nucleated and abundant food sources and elements and minerals necessary for growth were provided.

Likewise, as the zygot becomes compartmentalized and covered with a protective membrane and although consisting of up to 4 cell types, it does not grow in size. However, there follows rapid episodes of cell division creating up to 16 cells which ball together forming a morula and this is followed by another quantum leap in development and the morula grows larger in size and consisting of 128 cells; the clump of which is called the blastula which is studded with nuclei (Carlson 2013).

However, this clump of cells further differentiates into an outer cells, the trophoblasts which are not well differentiated, and inner cells which are well defined which becomes the embryo. The outer cells, the trophoblasts form a biosphere which provides nutrients and oxygen, and this is referred to as the ectoderm and which becomes the placenta (Carlson 2013).

Thus we see that an invasion of genes from one source, combines with the genes from a second source, generating a complex single cell then a multi-cellular organism, part of which forms a biosphere, the placenta, and the other part forming an embryo which is dependent on that placenta and which then becomes a fetus, then a neonate, and is then born into this world.

Thus, certain genes and nucleotide sequences are responsible for creating an embryo, others act to produce the placenta, and yet other genes or nucleotide sequences generate a fetus then a neonate. Once certain embryonic genes or gene segments have accomplished their function, they may trigger activation of different genes, or the next, adjacent nucleotide sequence segment, which produces yet new structures and tissues which overlie, extend, and add to those already formed. However, at the same time, cells are also dying off in massive numbers, due to programmed cell death; an example of which is the webbing that links the fingers but which disintegrates as these cells die off, and which could be likened to evolution followed by extinction as detailed in chapter 5.

Likewise, just as the placenta is jettisoned after it serves its purpose, innumerable species have died out once they had made their contributions to the next stage of "evolutionary" development. And once the "Eukaryote" is born it may then consume other Eukaryotes whose own genetic ancestry leads to the same ancestors. However, instead of splitting off to become placenta and embryo, they split off to be eaters and the eaten.

The parallels with the evolution of multi-cellular Eukaryotes, with microbes such as Cyanobacteria invading, providing genes, then biologically engineering the environment to sustain and promote Eukaryotic evolution and those events leading to extinction, are not exact, but nevertheless should not be dismissed as coincidence.

The mistake made by the Darwinists and neo-Darwinists is to assume that these complex interactions which have resulted in the evolution of the first multi-cellular Eukaryotes followed by increasingly complex and intelligent species are random events; which is roughly equivalent to assuming the fertilization of the ovum and the development of an embryo and placenta are random act of chance.

There are other parallels between embryogensis and evolution. Ontogeny does not replicate phylogeny. Nevertheless, the ancestry of primates leads to the tailless apes, then monkeys and finally to prosimian primates all of whom possess tails. Human embryos also grow a tail, which atrophies and disappears (Keith 2009; Laubichler & Maienschein 2007; Robert 2008). The ancestry of primates leads well beyond mammals to ocean dwelling animals who possessed gills which enabled them to breath oxygen. Human embryos also develop "gill slits" which undergo metamorphosis and become the pharyngeal arches and clefts of the throat, and which form the canals of the middle ear (Laubichler & Maienschein 2007; Robert 2008) as well as the parathyroid and thymus glands (Keith 2009; Robert 2008) which are involved in the metabolism of calcium. And calcium, like the oxygen breathed through gill slits and then lungs, owes its origins to Cyanobacteria.

Evolutionary Metamorphosis - Insects, Butterflies

Although "metamorphosis" is usually thought of as a transformation which occurs within a single season, the metamorphosis of a species may take millions or even billions of years. Metamorphosis in fact parallels evolutionary modes of punctuated equilibrium. During metamorphosis there are long periods of stasis, which are preceded or followed by quite sudden, abrupt and major changes in the animal's or insect's body structure (Duellman & Trueb 1994; Dofour et al., 2012, Gilbert 2009; Shi 1999; Wells 2007). Hence, the concept, "evolutionary metamorphosis" (Joseph 2000a).

A wide variety of species, including barnacles, bivalves, bryozoans, crabs, echinoderms, gastropods, mollusks, sea squirts, flat fish, flounder, eels, insects, amphibians, and reptiles, undergo profound physical alterations after birth and infancy which are often accompanied by massive cell death, the metamorphosis of new cells and body parts and significant changes in behavior and habitat (Duellman & Trueb 1994; Dufour, et al., 2012; Evans & Claborne 2005; Gilbert 2009; Heifman et al. 2009; Tackett & Tacket 2002). Flounder and other flatfish, for example, are born as a symmetrical fish larva, but then become decidedly asymmetrical and both eyes are displaced to the dorsal surface as they reach adulthood (Barss, et al. 1995; Evans & Claborne 2005; Gilbert 2009; Heifman et al. 2009; Tackett & Tacket 2002; Wells 2007). Eels undergo a tail to head metamorphosis with a complete shift of the digestive and urinary tracts from posterior to anterior.

Many species of insect first enter a larval followed by a pupal stage and then undergo transformations in morphology and physiology, growing wings, legs, reproductive organs and complex compound eyes which sprout from discrete enfolded pockets of tissue (Gilbert 2009). For example, and considered in the most generalized and broadest of terms, beginning with the fertilized egg the immature insect upon hatching rapidly grows into a larvae which can be up to two inches in length (Gilbert 2009). Caterpillars are larvae and they are incessantly active, their main activity being food consumption. And then, abruptly this frenzied activity comes to a sudden halt and is replaced by stasis, a long inactive stage referred to as chrysalis or pupa. The chrysalis/pupa stage is followed by a rapid series of transformations resulting in the metamorphosis of the adult, the caterpillar has become a butterfly. However, prior to and during that transformation, the multicellular creature that was a caterpillar also forms a biosphere and becomes a food source which nourishes what will become the butterfly (Gilbert 2009).

In a variety of insects as they undergo seasonal metamorphosis, there are flurries of genetic activity, with genes being turned on and off, whereas other genes are being duplicated and expressed. This results in a cellular and anatomical metamorphosis, such that body parts, and their cellular constituents, are transformed in accordance with the DNA instructions transmitted. Cells and body parts may be broken down and reassembled and new cells manufactured in their place. The caterpillar is completely transformed into what *appears* to be a wholly different species: a butterfly.

Molting and Metamorphosis

Anthropods can also shed one body in place of another, a metamorphosis that involves "molting" (Gilbert 2009). It is in this manner, for example, that arthropods are provided the opportunity to "evolve" into a completely different body with different behaviors, albeit over a single season.

Because of their hard outer cuticle, many species of insect are forced to molt in order to grow. Molting and metamorphosis require the differential activation and suppression of various genes and nucleotide sequences, and/or the coding of a greater or shorter length of the same segment of base pair sequences.

For example, during metamorphosis, and in response to various internal and external signals, insect DNA located in brain cells, begin to manufacture and induce the secretion of enzymes which acts on the cellular DNA of the thoracic glands which in turn produces and secretes a molting hormone, thyroxin (Gilbert 2009). Thyroxin is an iodine containing hormone which is also secreted by all vertebrates and promotes development and physical maturation in humans. In insects, the secretion of this and related hormones (e.g. ecdysone) acts on the DNA of the epidermal cells (via intracellular protein receptors) causing them to detach from the old cuticle which is simultaneously being dissolved by the DNA secretions of yet another cell.

In caterpillars, over 90% of the original cuticle is digested by these enzymes (Gilbert 2009). This process enables growth and involves the DNA of one cell acting on itself and the DNA of another cell within the same organism; which not only involves programmed cell death but, in the case of butterflies, the transformation of cells which become food sources for those cells which will become the butterfly.

Following molting, metamorphosis involves differential DNA activation and then the secretion of yet other enzymes including "juvenile" hormones, e.g. farnesol which is produced by the DNA located in the corpus allatum gland. (Gilbert 2009). In fact, the same hormones which induce insect metamorphosis, farnesol and ecdysone, are found in almost all Eukaryotic species, including humans and are produced by the same genes. ,

These DNA produced hormones also enable the insect to enter into a second embryonic, or larval stage. In consequence, each and every cell begins to undergo a transformation as old cells are genetically reengineered or discarded. The insect's head may emerge from what had been a tail, and wings emerge from thoracic muscles; and what might have taken five hundred million years to accomplish, takes place in a single season.

However, if that same metamorphosis actually took five hundred million years, instead of a single season, the Darwinians would claim that the appearance of genetic design is an "illusion," that the *evolution* of the butterfly is not under genetic regulatory control, but is due to the natural selection of chance and minor variations including "random mutations"

Metamorphosis vs Evolution: Frogs

Amphibian metamorphosis is yet another dramatic example which parallels basic principle of evolution, and involves a series of major physical transformations as these animals grow from egg to a larva/tadpole to an adult (Duellman & Trueb 1994; Wells 2007). Frogs may have first evolved around 265 million years ago, but their development and metamorphosis parallels the evolutionary concept of punctuated equilibrium.

Frog eggs are usually deposited in water and will often clump together (Shi 1999). Following basic principles of embryogenesis and Eukaryogenesis including the biological engineering of the biosphere, part of the frog embryo becomes a food source for the embryo; a jelly that also acts as a protective membrane which allows oxygen and other nutrients to enter (Wells 2007). The developing embryo then secretes various enzymes which split open the egg, and a fish-like tadpole emerges, equipped with gills and a tail which enables it to swim, fish-like, in the water.

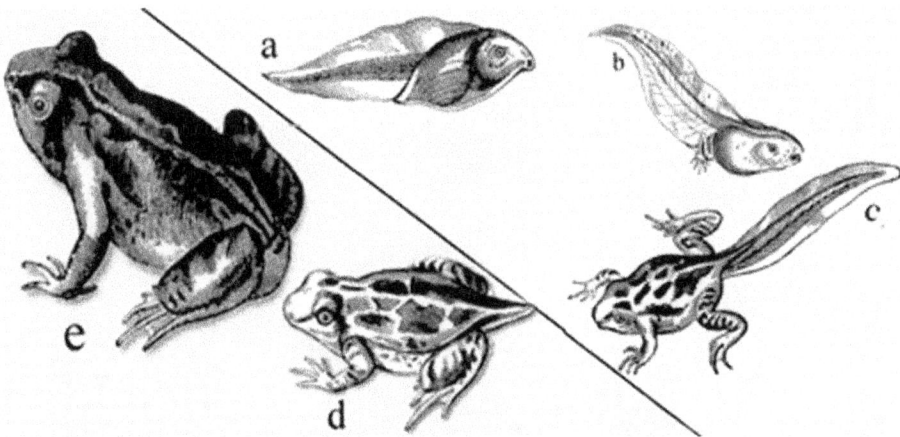

The tadpole remains in stasis and then sprouts hind legs and front legs whereas the cells making up its tail disintegrates and is absorbed by the body serving as source of nutrients. The gills are replaced by lungs, eyes migrate rostrally and dorsally giving it stereoscopic vision, and its skin and blood supply undergoes a chemical transformation. In addition, its mouth, intestines, and digestive tract become transformed and these animals become carnivores and thus change their diet, mode of digestion, and excretions from nitrogen and ammonia to urea (Shi 1999; Wells 2007). Moreover, they grow an additional brain structure, the cerebellum, which sits atop the brainstem, and which coordinates gross body movement including maintaining stability when leaping and jumping. Thus, the skin, eyes, mouth, internal organs, skeletal-muscular system, and the brain, undergo dramatic and major transformations, and the fish-like tadpole becomes a frog who no longer lives in the water, but near the water on land (Duellman & Trueb 1994; Shi 1999; Wells 2007).

Thus, again, we see dramatic alterations, where one type of animal becomes a wholly different animal; a process which is completely under genetic and environmental control, but which would be called "evolution" if it took place over a hundred millions years and not just in a few days.

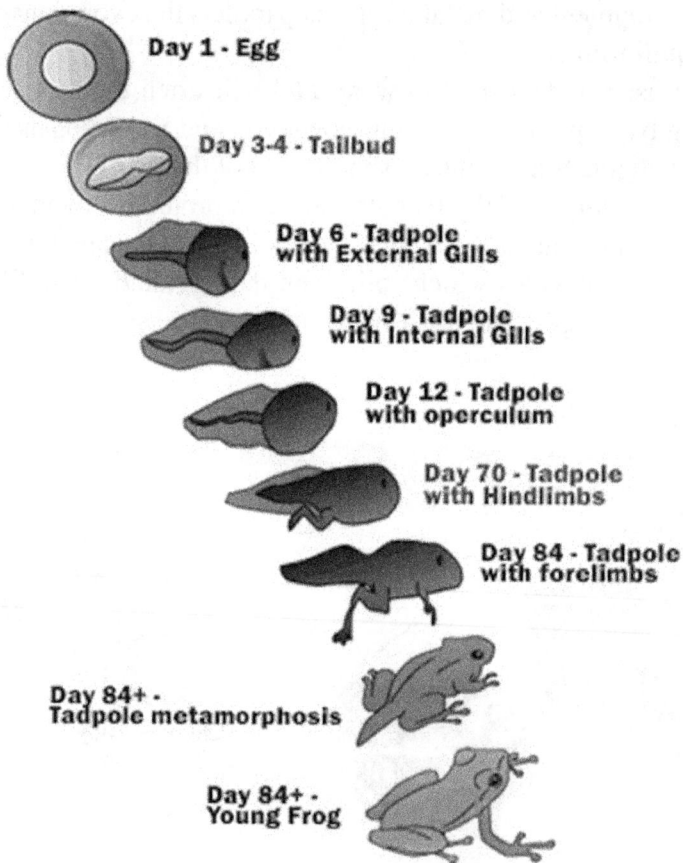

Day 1 - Egg

Day 3-4 - Tailbud

Day 6 - Tadpole with External Gills

Day 9 - Tadpole with Internal Gills

Day 12 - Tadpole with operculum

Day 70 - Tadpole with Hindlimbs

Day 84 - Tadpole with forelimbs

Day 84+ - Tadpole metamorphosis

Day 84+ - Young Frog

DNA-Supra Organism

A secondary premise of the theory of evolutionary metamorphosis and evolutionary embryology, is that all forms of life are linked at the level of DNA, thereby creating an evolving supra-DNA-organism consisting of multiple genomes and genes which act collectively to promote the continued evolution of the supra-organism. Therefore, just as sperm attaches itself to an egg and there is an exchange of DNA which results in the generation of a multi-cellular zygot, bacteria, archae and viruses combined genes to generate multi-cellular Eukaryotes and also invaded these Eukaryotes.

Moreover, just as occurs during metamorphosis and embryogenesis where portions of the larvae or zygot split off to become food sources and create a biosphere for the larvae and zygot, Eukaryotes consume Eukaryotes whereas Prokaryotes and later some species of Eukaryotes biologically manufactured a biosphere.

For example, depending on species, insect metamorphosis often involves complex gene-environmental interactions resulting in the death of most of the cells of the body, referred to as holometabolism. Holometabolism can be considered a form of cell-suicide. The insect's body will "digest" itself and will feed upon its dead cells and then rapidly grow, undergoing a radical transformation into a seemingly completely different species. Similar events have taken place over the course of evolution, with numerous microbial species dying following dramatic alterations in the world climate and then becoming food sources for those yet to evolve. Likewise, Eukaryotes consume Eukaryotes and all presumably split off from the same ancestors.

Evolution can be likened to embryology and metamorphosis. Moreover, the DNA of diverse species are interactional and together may be viewed as constituting a supra-DNA-organism which acts on the environment in order to promote DNA activation and dispersal and the metamorphosis and continued development of this supra-organism. That is, DNA acts to genetically engineer the environment which in turn acts on gene selection which acts on the environment which leads to the expression of those genes inserted into the Eukaryotic genome by Prokaryotes and viruses.

Likewise, just as DNA contains the instructions for creating and nourishing an embryo in parallel with the genetic alteration of the placenta and the womb, DNA also contains the instructions for altering itself and the external environment so as to promote not just diversity, but the emergence of increasingly complex and intelligence animals, including the likes of woman and man.

Recognizing the role played by diverse species-and thus their DNA in the biological construction of the atmosphere, climate, and even the chemical, mineral and oxygen contents of the oceans, is integral to understanding evolutionary metamorphosis, including extinction and evolvability vs the failure of some species to "evolve." Just as body parts or dead cells may be sloughed off or absorbed during metamorphosis and embryonic, fetal, and neonatal, development, over the course of "evolution" some species have been sloughed off and become extinct once they were no longer needed. Just as some body parts remain relatively simple--the cells of the skin versus the nerve cells of the brain--over the course of "evolution" some species have remain relatively simple and others have become more complex--all are integral to the survival of the supra-DNA-organism.

Therefore, the DNA or diverse species can be seen as an interactional supra-DNA organism which acts on the environment by altering the biosphere or serving as food sources for itself. Once certain organisms (*branches and twigs from the tree of life*) have accomplished their genetic mission, they become extinct, whereas other cells, including simple organisms continue to thrive as their output is essential to maintaining and promoting life; that is, the life of the supra-DNA-organism. Coupled with unforeseen environmental catastrophes, the need to modify the environment in order to promote DNA dispersal and development, and the fact that the environment acts on gene selection and activation explains the periodic lack of progress in complexity over eons of time, and then the sudden surges in progress and complexity, during different epochs of the Earth's history.

However, what this also demonstrates is that the "genetic seeds of life" which swarm throughout the cosmos, contained the genes, genomes, and the genetic instructions for the evolution and metamorphosis of every creature which has walked, crawled, swam, or slithered across the Earth, including woman and man.

Embryogenesis and metamorphosis are under precise genetic regulatory control, and so too is what has been called: *Evolution. Evolution* is a complex process of embryological metamorphosis; the germination of the *genetic seeds of life* which swarm throughout the cosmos.

8. Multi-Regional Human Metamorphosis

The Ticking of the Genetic Clock

The progressive metamorphosis of increasingly intelligent and complex life on this planet, culminating in the emergence of modern woman and man, appears to have unfolded in a genetically predetermined, "molecular clock-wise" fashion. That "evolution" is regulated in accordance with the "ticking" of a genetic "clock" is evident from an examination and comparison of various genes and ribosomal RNA belonging to diverse species. Because this genetic "clock" appears to have been "ticking" at the same rate, simultaneously, among all branches of the tree of life (Denton, 1998), and as this "clock" on Earth, began to "tick" almost 4 billion years ago, it thus appears that the continued "ticking" of this same "genetic clock" has determined the successive emergence of increasingly intelligent and complex species (e.g., Kumar & Hedges, 1998), culminating in woman and man.

Just as the genome of the caterpillar is programmed to produce a butterfly, or the DNA of a fertilized ovum gives rise to an embryo then a fetus... neonate... child... juvenile... adult, the metamorphosis and progression which characterizes life on this planet appears to have been preprogrammed into the DNA of some of the first Earthlings. It is this genetic predetermination which also explains why although the Earth has been five times struck by life-destroying meteors, that certain species immediately recovered, and why no new phyla have emerged since the Cambrian Explosion.

Because the emergence of H. sapiens sapiens, and all manner of species appears to have been genetically preprogrammed, this is referred to as "evolutionary metamorphosis" (Joseph, 1997, 2000a). However, rather than a 9-month gestation period, or the single seasonal metamorphosis which characterizes the transition from caterpillar to butterfly, humans are an end product, or perhaps a mid-way product of a process which takes several billion years to unfold; albeit in accordance and in parallel with suitable changes in the genetically engineered environment.

Thus, metamorphosis is not a one-step progression (caterpillar-butterfly) but a leaping, branching, multi-step progression involving numerous successive species, many of which are genetically preprogrammed to give rise to the next in a "molecular clock-like" fashion. The ticking of this genetic clock, however, also requires that the environment be genetically engineered in preparation for those yet to be born (Joseph, 1997).

127

DNA-Supra Organisms

A premise of the theory of evolutionary metamorphosis is that microbes and in fact all forms of life consist of packages of DNA which have manufactured an organism also interact with the environment. A fat hairy spider crawling along the ceiling is the product of DNA engineering, and each and every spider-cell contains a packet of DNA which created the cell and which contributed to the creation of the spider. The "spider" is merely a vehicle through which the DNA navigates its way around the world. The same is true of fish, frogs, reptiles, and so on. These are manifestations of DNA activity and every organism functions in accordance with specific DNA-instructions.

DNA strives for expression and dispersal. According to the theory of evolutionary metamorphosis, the DNA of diverse species are also interactional and together may be viewed as constituting a supra-DNA-organism which acts on the environment in order to promote DNA activation and dispersal. That is, DNA acts to genetically engineer the environment which in turn acts on gene selection which acts on the environment which leads to the expression and dispersal of additional DNA. Thus, just as DNA contains the instructions for creating and nourishing an embryo in parallel with the genetic alteration of the womb, DNA also contains the instructions for altering itself and the external environment so as to promote not just diversity, but the emergence of increasingly complex and intelligence animals, including the likes of woman and man.

Recognizing the role played by diverse species-and thus their DNA-- in the biological construction of the atmosphere, climate, and even the contents of the oceans, is integral to understanding evolutionary metamorphosis, including extinction, and the failure of some species to "evolve." Just as body parts or dead cells may be sloughed off or absorbed during embryonic, fetal, and neonatal, development (Joseph, 2000b), over the course of "evolution" some species have been sloughed off and become extinct once they were no longer needed. Just as some body parts remain relatively simple, e.g. the cells of the skin versus the nerve cells of the brain, over the course of "evolution" some species have remain relatively simple and others have become more complex--all are integral to the survival of the supra-DNA-organism.

Again, the DNA or diverse species can be seen as an interactional supra-DNA organism which acts on the environment. Once certain organisms have accomplished their genetic mission, they become extinct, whereas other cells, including simple organisms continue to thrive as their output is essential to maintaining and promoting life--that is, the life of the supra-DNA-organism. Coupled with unforeseen environmental catastrophes, the need to modify the environment in order to promote DNA dispersal and development, and the fact that the environment acts on gene selection and activation (e.g., de Jong & Scharloo, 1976; Dykhuizen & Hart, 1980; Gibson & Hogness, 1996; Polaczyk et al., 1998; Rutherford & Lindquist, 1998; Wade et al., 1997) explains the periodic lack of progress in complexity over eons of time, and then the sudden surges in progress and complexity, during different epochs of the Earth's history.

Genetic Engineering and Stabilization of the Environment: Simplicity

It is necessary that some species remain relatively simple and basically identical to their ancestors from billions of years ago. There is a genetic need to maintain certain environmental, climatic and atmospheric conditions (such as oxygen secretion). In consequence, certain species never progress. In fact, there are specific repressor proteins and a variety of genetic mechanisms which act to prevent genetic change, even in response to changing environmental conditions. For example, regulator proteins referred to as "chaperones have been found in all organisms studied and protect against" genetic change or activation such as in response to changing environmental and climatic conditions and other stresses, such as alterations in oxygen levels (Cossins, 1998).

For example, a genetically manufactured protein, "Hsp90 is one of the more abundant chaperones. At normal temperatures it binds to a specific set of proteins, most of which regulate cellular proliferation and embryonic development. These signaling proteins form complex webs of molecular switches that allows signals both within and between cells to be transduced into responses... and act against genetic variation" and prevent the expression of silent characteristics (Cossins, 1998, pp. 309-310). For example, these proteins may prevent DNA expression by acting as a buffer between these silent genes and nucleotides and the environment, so that they are not expressed except in accordance with specific genetic instructions.

Again, consider Hsp90. Hsp90 targets multiple signal transducers which control and act as "molecular switches" which in turn control gene expression. Hsp90 "normally suppresses the expression of genetic variation affecting many developmental pathways" (Rutherford & Lindquist, 1998). However, Hsp90, also reacts to environmental stress including diet and fluctuations in temperature (Rutherford & Lindquist, 1998). Indeed, as demonstrated by, Rutherford and Lindquist (1998, p. 341) Hsp90 acts as an "explicit molecular mechanism that assists the process of evolutionary change in response to the environment" and it accomplishes this through the "conditional release of stores of hidden morphological variation.... perhaps allowing for the rapid morphological radiations that are found in the fossil record."

However, in order for these repressor proteins and other regulating genetic mechanisms to be switched off or on, requires contact and exposure to specific environmental agents.

Initially, the new Earth was devoid and lacking these environmental agents, such as free oxygen, calcium, and so on. Hence, in order for certain genes and gene sequences to be activated, required that these products be liberated and/or manufactured. Hence, some species immediately began secreting oxygen as a waste product, which in turn acted on gene selection.

It is because of the genetic need to create a precursor product in massive amounts (such as calcium), which explains why some species emerge, thrive, alter the environment, and then become suddenly extinct. Some species emerge simply to produce a specific product and are then jettisoned. Just as the placenta is a nurturing biological construction that is jettisoned with the birth of the baby, there have been periods when much of the Earth's biomass served only to produce and secrete products that were fundamental for the metamorphosis of those yet to be born, such as calcium to build bones (Joseph, 1997). Once their genetic mission was accomplished, many of these creatures were jettisoned and became extinct.

Consider, again, the calcium carbonate secreting Ediacaran fauna. As oxygen levels increased in the atmosphere and in the sea, and as the planet again began to warm, oxygen breathing multi-cellular eukaryotes emerged (e.g. Brocks et al., 1999). By 2.3 billion years ago the Earth's land masses were covered with thick bacterial mats and other organisms. Moreover, many of these organisms secreted a variety of organic acids which in turn formed laterites (iron rich deposits) by leaching iron from the upper layers of rock and soil. However, as pointed out by Dr. Ohmoto, "in order for laterites to form, there must be organic material and atmospheric oxygen," substances secreted by and the residue of even earlier life forms.

By 1.6 billion years BP., the Earth's climate and environment had been dramatically altered and animals began evolving into different species (e.g., Hedges & Kumar, 1999) who in turn began to prepare the world for subsequent generations. Then around 600 million years ago the calcium carbonate secreting Ediacaran emerged in every ocean and sea, releasing materials into the environment which enabled shellfish and bony complex oxygen-breathing creatures to "evolve" and undergo metamorphosis.

However, the Ediacaran fauna and subsequent generations would not have been able to thrive if not for the thick layers of bacteria which had been building up for over 2 billion years --much of which then served to nourish the Ediacaran fauna and those who emerged during the early phases of the Cambrian Explosion--just as thick mats of blood cells sustain the ovum within the womb. One species served as the nutrients for a later appearing species who prepared the world for the next generation of increasingly complex organisms.

The Metamorphosis Of The Ediacaran Fauna

Ediacaran fossils have been discovered throughout the world, and date from 580 billion to 560 billion years BP. (though some authors have assigned them a date of 600 million years BP.) These were soft bodied, leaf- and disk-shaped, plant-like creatures, consisting of only 11 or fewer cell types (compared with over 200 cell types for mammals). They ranged in size from over 3.5 feet to less than 1/2 inch (Glaessner, et al. 1988) and then suddenly became extinct.

The emergence of Ediacaran fauna was not just a genetic experiment gone awry, as some scientists have speculated, for these creatures and their waste products altered the planet so as to make the next stage of metamorphosis possible. Because they secreted calcium carbonate, from which shells and bones are constructed, the Ediacaran fauna made possible the metamorphosis of shell fish and the skeletal system. Once their genetic mission was accomplished, the Ediacarans disappeared from the scene and there followed an explosion of life in every ocean, lake, river and stream--aptly referred to as the Cambrian Explosion as it took place during the Cambrian era.

The Cambrian Explosion

"If it could be demonstrated that any complex organ existed which could not possibly have been formed by numerous successive. slight modifications, my theory would absolutely break down." -Darwin, 1857.

With no history of derivative ancestral forms, all manner of complex life forms suddenly emerged with gills, intestines, joints, brains, and modern eyes equipped with retinas and fully modern optic lenses. These included organisms with a hard tube-like outer-skeleton consisting of calcium carbonate, and all manner of "small shelly fish" (Anabrites, Protohertzina), as well as sponges, jelly fish, mollusks, brachiopods, and the first arthropods (e.g. trilobites) which immediately sprouted legs and primitive brains. In fact, every phylum in existence today emerged during the Cambrian Explosion, including some phyla which emerged then became extinct.

The survivors, however, included the phylum Chordata; i.e. tunicates and the first jawless fish who possessed a notochord and simplified brain that consisted of a brainstem and limbic forebrain.

Hence, during the Cambrian epoch there was also a cerebral and thus a cognitive explosion as the first true brains were established, brain which would continue to undergo a genetically preprogrammed metamorphosis until finally ending up in human heads.

And yet, just as the Ediacaran fauna emerged, secreted massive amounts of calcium, and then departed the scene, an incredible number of phyla emerged during the Cambrian explosion, made their own genetically engineered contributions, and then became extinct--a pattern of sometimes inexplicable species extinction that has been repeated time and again. Again, just as the placenta is jettisoned after it serves its purpose, innumerable species have died out once they had made their contributions to the next stage of "evolutionary" development.

Thus, we see that major species have emerged, flourished, diversified, modified the environment, and then died out, only to be followed by yet another wave of "novel" species who followed the same pattern, releasing and secreting additional "waste" products and thus making their own contribution to the environment and the development of the next wave of diverse species, their descendants, before disappearing from the scene. However with each successive wave there has been not just diversity, but progress, environmental change, and increased complexity in design and intelligence, leading from simple cellular organisms to all manner of species, including woman and man.

Evolutionary Metamorphosis

The astrobiological evolutionary metamorphosis theory of life is a complete departure from Darwin's theory and neo-Darwinian theories of evolution, which instead claim that "evolution" is due to random mutations and random variations. It is also a complete departure from those more extreme neo-Darwinians who deny progress and that evolution is characterized by the emergence of increasingly complex and intelligent animal life.

The theory of evolutionary metamorphosis is based on genetics and the fossil record. By contrast, Darwin's theory is not supported by the fossil record and it is refuted by genetics--as there is nothing random about DNA/RNA or DNA/RNA expression--the source of all supposed "random mutations." Indeed, his theory does not even taken into account progress, but only variation; and aspects of life on this planet has become not just variable, but increasingly complex.

As repeatedly stressed, "traits" exist prior to their expression, being genetically preprogrammed into the genetic code (de Jong & Scharloo, 1976; Dykhuizen & Hart, 1980; Gibson & Hogness, 1996; Polaczyk et al., 1998; Rutherford & Lindquist, 1998; Wade et al., 1997). And, not just traits, but the instructions for the creation of increasingly complex species--and progress and the fossil record is not compatible with Darwin's theory.

What is supported by the fossil record, is the theory of "natural selection" which was first developed by A.R. Wallace. Wallace's theory of "natural selection" (and particularly his theory as to the evolution of woman and man) is fully compatible with the theory of evolutionary metamorphosis.

With the exception of present day "creation scientists" most scientists agree with the notion that the evolution of almost all Earthly life trace their ancestry to those creatures who were among the first to emerge on this planet. However, Darwin and neoDarwinians champion the "organic soup" and believe that life and its DNA emerged from non-life following the random mixing of organic molecules. And yet, there was little or no free oxygen, which is necessary for the construction of DNA. The basic elements for DNA construction did not exist on Earth.

However, let us pretend that life did emerge from an organic soup. If that is the case, then, given the lack of Earthly ingredients, that soup must have first been stirred on another planet.

Cosmic collisions are commonplace, not only between stars, but entire galaxies. And, if life first arose on another planet, it can be assumed that some of those creatures were cast into space, encapsulated in debris, and not only survived their long journeys, but fell upon innumerable planets, only to begin genetically engineering their new worlds if at all possible.

Hence, although life may ultimately trace its roots to a single cell that emerged from an organic soup, the theory of evolutionary metamorphosis views life as having arisen, on Earth, from numerous "seeds" which, in turn, may ultimately trace their origin to a single extraterrestrial source.

If there was a "single seed" from which all life has descended, this "seed" first appeared tens of billions if not hundreds of billions of years ago, on another planet thus giving rise to identical seeds; that is creatures with identical or similar DNA and DNA-based genetic instructions.

Because of this genetic commonalty, rather than a single seed and a single Earthly-trunk with innumerable branches, the theory of evolutionary metamorphosis posits a forest of trees with innumerable branches, each of which has the genetic potential to bear identical fruit. Only the theory of evolutionary metamorphosis and this forest of basically identical "genetic trees" can explain why different species of humanity, such as Homo erectus, Neanderthals, and Cro-Magnons shared the planet during overlapping time periods

According to the theory of evolutionary metamorphosis, since the genetic instructions for creating all manner of life is also DNA based, then it would appear that these genetic instructions and the genetic potential to create all manner of life, are also astrobiological in origin. It is these ancestral astrobiological origins, the antiquity of life, and the genetic memories and instructions which they have passed down, which explains the progressive emergence of increasingly complex and intelligent life on Earth--an unfolding which has occurred according to specific DNA-based instructions and genetic memories.

Once on Earth, the descendants of these astrobiological life forms, and their DNA, began to multiple, to genetically engineer this planet, thereby altering the environment which acted on gene selection, thus giving rise to increasingly intelligent and complex species; albeit in accordance with genetically predetermined instructions--just as the fertilized single cell that gives rise to an embryo is genetically predetermined to create a fetus, then a neonate, child...adult.

Just as an embryo is not a random construction, all subsequent species to emerge on Earth have been genetically preprogrammed, the expression and coding of which are associated with intronic genes and intronic gene sequence activation, exon shuffling, "frame shifts," intron and plasmid insertion and exchange, and a host of other genetic variables which are yet to be identified. Once expressed, these genes act to genetically engineer the planet, and to modify the environment which acts on gene selection, which acts on the environment, which acts on gene selection, and so on.

As repeatedly noted, it has now been demonstrated that "populations contain a surprising amount of unexpressed genetic variation that is capable of affecting certain" supposedly "invariant traits" and that changes in environmental conditions "can uncover this previously silent variation" (Rutherford & Lindquist, 1998 p. 341). That is, traits may be expressed and which exist prior to their expression as they are predetermined and precoded within "silent" genes and "silent" nucleotide sequences. Consider, over 96% of the human genome is "silent" and the genomes of most eukaryotes contain vast amounts of silent genes.

Moreover, as has been demonstrated experimentally, these silent genes, and these silent traits, can be expressed by varying the environment and through other stresses including fluctuations in temperature, oxygen levels, and diet (de Jong & Scharloo, 1976; Dykhuizen & Hart, 1980; Gibson & Hogness, 1996; Polaczyk et al., 1998; Rutherford & Lindquist, 1998; Wade et al., 1997).

The environment acts on gene selection--genes which exist and which contain unexpressed traits that also exist prior to their expression but which are repressed by repressor proteins which coat these genes/nucleotides and thus prevent their activation or recognition by RNA. These silent genes are also regulated by other genes and gene protein products, including, as noted above, Hsp90 (Cossins, 1998; Rutherford & Lindquist, 1998, p. 341), which acts as an "explicit molecular mechanism that assists the process of evolutionary change in response to the environment" and it accomplishes this through the "conditional release of stores of hidden morphological variation.... perhaps allowing for the rapid morphological radiations that are found in the fossil record."

Silent genes make up the majority of the genome--at least for most species, including humans--and are considered a "hot spot" for homologous recombination (Wahls et al. 1990). Specifically, introns are responsible for producing and giving birth to "new" genes and "new" gene clusters, and to duplicate genes and gene families, including numerous copies of highly repetitive sequences of nucleotide base pairs (Finnegan, 1989; Henikoff et al. 1986; Petcs & Fink, 1982); the creation of which is not due to random factors but is under precise genetic control. The result is the creation of "new" genes-- genes which had been genetically preprogrammed (Joseph, 1997)-- and the creation of numerous sequence families (or gene families), some of which are dispersed throughout the genome (Jelinek et al. 1980) and the expression of new products and traits which had been coded into these silent genes prior to their expression (de Jong & Scharloo, 1976; Dykhuizen & Hart, 1980; Gibson & Hogness, 1996; Polaczyk et al., 1998; Rutherford & Lindquist, 1998; Wade et al., 1997).

The existence of these silent genes and genes within genes, and traits which are repressed but require only the right environmental signal to emerge, explains why evolution is not random, and why the emergence of seemingly new species is under the precise control of genetic "molecular clocks."

The gradual unlocking of this intronic genetic potential and these genetic memories, has therefore resulted in a linear, branching, sequential, "molecular clock-like" unfolding that involves the activation of ancestral genetic memories.

Only the purposeful expression of genetically pre-coded instructions can account for the obvious evidence of a step-wise, sometimes leaping progression in increasing intelligence and complexity which has characterized the metamorphosis of a rather narrow range of life on this planet. Indeed, only precise genetic instructions can account for the fact that basically similar species have emerged multi-regionally across distant lands, from distinct pockets of ancestral species which also emerged multi-regionally, and this includes the multi-regional metamorphosis of human beings such as Homo erectus, Neanderthals, and Cro-Magons. The planet was genetically seeded to grow all manner of species, including humans and all manner of variations thereof.

Evolutionary Progress And Multi-Regional Metamorphosis

Every phylum in existence today emerged during the Cambrian Explosion, including the phylum Chordata. The first members of the phylum chordata possessed a simplified brain that consisted of a brainstem--which controls rhythmic and reflexive motor behaviors-- and a limbic forebrain which mediates all aspects of emotional and motivational functioning including memory.

The evolution of the brain began with the metamorphosis of the first nerve cells; i.e. specialized sensory-motor cells capable of inducing movement in reaction to sensation such as light vs shadow. These cells were loosely organized along the outer membrane/skin and most probably did not intercommunicate except indirectly following the release of various chemical transmitters.

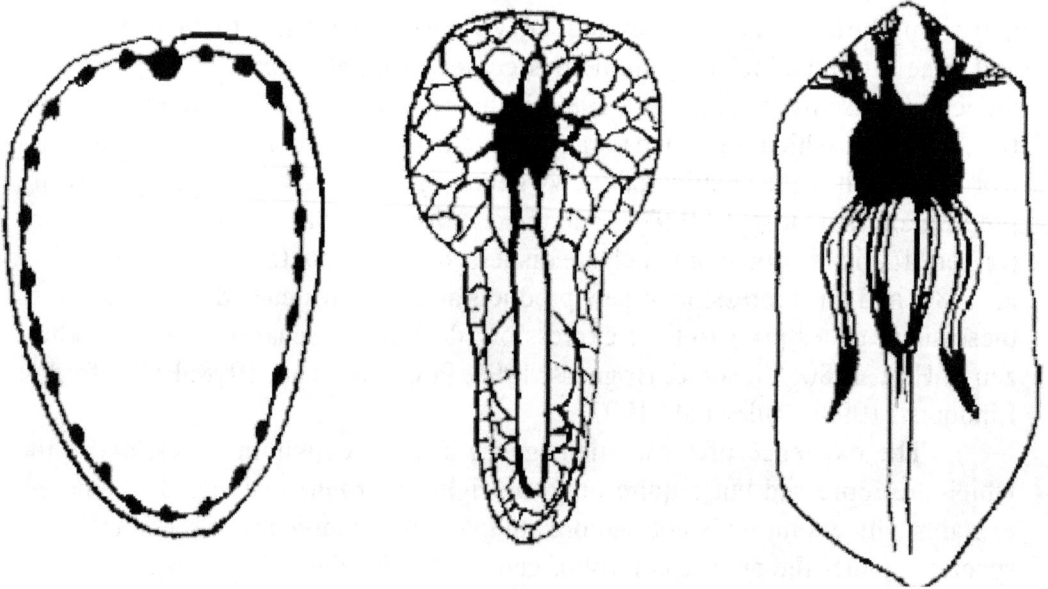

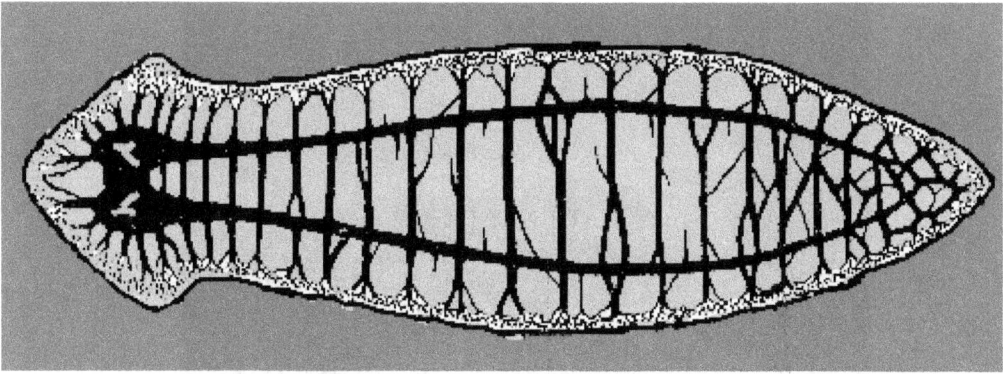

As the climate and environment began to change, and as the environment acts on gene selection, over the course of evolution a collection of like-minded cells began to directly interact, forming a nerve net, and then to congregate in the anterior head region, giving rise to a primitive ganglion brain. During the Cambrian Explosion, the ganglionic brain became a primitive brain, which in some species including cartilaginous fish (e.g. sharks) and later, in "bony fish" (Osteichthyes), consisted of a brainstem (concerned with reflexive sensory motor functions) and an olfactory bulb-equipped forebrain which analyzed environmental chemical input and also induced gross motor behavior in response to motivationally significant stimuli.

Tunicates were among the first chordates (subphylum Urochordata) to emerge beneath the sea, some 500 million years ago, and were soon followed by the first jawless fish (e.g. Astraspis, Arandaspis), who in turn gave rise to cartilaginous (Cyclostomes) "bony" fish. Over the ensuing 100 million years, and within the vast oceans and seas, various species of "fish," e.g. "bony fishes," "ray fins," developed additional brain matter, and some species of "bony fish" later developed lungs and limbs; i.e. lobed finned fish.

Armored and jawless fish, sharks, and Lobed finned fish were all in possession of the prototypical brain, the basic framework of which would be inherited by all subsequent species, including amphibians, reptiles, and even woman and man (Nieuwenhuys & Meek, 1990b; Stephan, 1983).

The brains and bodies of these animals, however, did not become just more variable--as demanded by Darwinian theory--but increasingly complex and hierarchically organized and sophisticated. For example, unlike other fish which are externally fertilized and which lay eggs in the open water (eggs which are then greedily gobbled by yet other denizens of the sea), the lobe finned fish were fertilized inside the body and could bear the young alive.

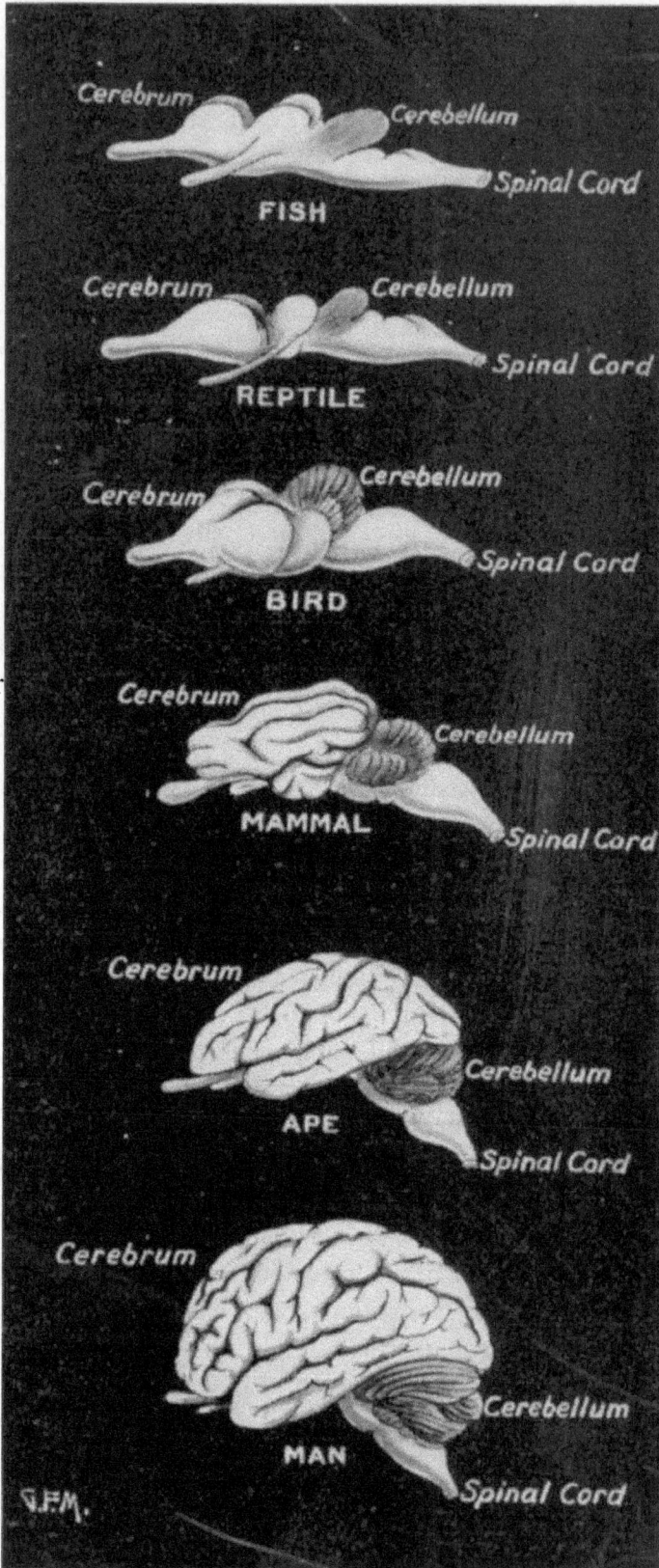

By 370 million years B.P., a wide range of lobed finned fish began to appear in almost every ocean and sea. These included Dipnoans, Sarcopterygia, and Coelacanths, who began to venture forth upon the Earth where they then began to breed.

Numerous species of lobed-fins lived mainly in rivers and freshwater seas and could venture forth and live on land as they had evolved internal air sacs which were embedded within their fins. Likewise, some of the first land-based plants also evolved air sacs. These air sacs could pass oxygen directly into the blood stream. This "breathing" ability enabled the lobed fins not only to venture forth, but to hole up in caked mud during the dry seasons. As the environment acts on gene selection, it is the lobed finned fish who presumably gave rise to the next stage of animals life.

Lobe finned fish were (and are) in many respects a transitional prototype for all land based creatures, as their "fleshy-lobed" fins were supported by an internal skeleton consisting of a humerus, femur, radius, ulna, tibia and fibula (Caroll, 1988; Jerison, 1973; Nieuwenhuys & Meek, 1990b; Romer, 1970). The lobe finned Coelacanths, in fact possessed jointed bones shaped somewhat like arms and legs.   It is from these lobed fins that legs would eventually "evolve" (Caroll, 1988; Colbert, 1980; Jarvik, 1980; Jerison, 1973; Romer, 1970), and it is these lobed fins coupled with the air sacks (primitive lungs), that enabled these creatures to periodically leave the water so as to venture along river banks, oceans fronts, and onto dry soil, some 400 to 350 million years ago (Caroll, 1988; Colbert, 1980; Jarvik, 1980; Jerison, 1973; Romer, 1970).

Presumably it is from one or any number of the various species of lung/air sack equipped lobe finned fish, that all terrestrial vertebrates evolved (Romer, 1970), beginning, perhaps with primitive oxygen breathing amphibians, such as the seven fingered Ichthyostega whose forelimbs were hitched to the skull.

This odd physical organization, however, enabled these and like-minded animals not only to walk but to perceive and hear vibration transmitted through their feet.

By 360 million years ago a variety of five-toed amphibians, some up to 15 feet long were swarming over the planet. And, for a brief time amphibians "ruled" the world as they were more social, and more intelligent than the more solitary insects who, along with planets, had dominated the planet. Insects rapidly diminished in size.

It was not the lobed finned lung fish, however, but their descendants that gave rise to amphibians. These amphibian-like creatures looked something like a cross between a fish and a big salamander, with flat heads and long tails, and short stocky feet like a turtle. These include the eusthenopterons, as well as the Ichthyostegas which used four feet in order to move about (Colbert, 1980; Jarvik, 1980; Jerison, 1973; Romer, 1970).

Hence, by 350 million years ago the lobe finned fish presumably evolved into a fish with legs, the eusthenopteron and ichthyostegas, which in turn evolved into amphibians, some of which grew up to 15 feet length and who sported an enlarged olfactory lobe which dominated the forebrain.

Reptilian Metamorphosis

Amphibian dominion was sooner overturned by reptiles who were superiorally endowed physically and neurologically as their forebrain consisted of a greatly expanded limbic system and included distinctive limbic system structures such as the amygdala, hippocampus, and striatum which conferred tremendous intellectual powers and social-emotional signaling and associational capabilities upon these animals. Amphibians, therefore, did not just become more variable, they gave rise to a unique and superior species.

Unlike their amphibious cousins, the reptiles were better engineered for living on dry land, having evolved scaly water-proofed skins, hip and shoulder girdles, as well as a "new" method of giving birth. In contrast to amphibians who must return to the water to breed and produce young (tadpoles which undergo metamorphosis in order to become an adult), the reptiles could breed and lay amniote/cleidoic (shell covered) eggs on land from which emerged miniature adults.

The brain also increased in size and became more complex. With the evolution of reptiles the limbic forebrain mushroomed in size and gained hierarchical control over the motor functions of the brainstem (Herrick, 1948; Nieuwenhuys, 1967). The limbic forebrain could now feel emotions and directly control all aspects of body movement. In part, the increase in the size of the forebrain was induced by the increased emotional and motor demands of living on dry land, such that the motor aspects of the forebrain began to increasingly differentiate and to evolve in response to and in order to meet these new motoric needs.

Yet another factor in the encephalization of the brain was that animals were now living in a perfumed world of smell, and these odors provided an incredible wealth of information that the limbic forebrain became specialized to analyze. As the environment acts on gene selection, olfactory input to the limbic system forced this structure to also evolve and differentiate and to become increasingly capable of analyzing a wide range of motivational stimuli. The brain did not become more variable, as demanded by Darwin's theory, but larger and increasingly complex.

Therapsid & Repto-Mammal Metamorphosis

Over the ensuing 25 million years, the descendants of reptiles diverged, one branch of which giving rise to the intellectually, neurologically, and physically advanced repto-mammals, around 250 million years B.P., who in turn gave rise to therapsids who emerged multi-regionally (Bakke, 1971; Brink, 1956; Crompton & Jenkins, 1973; Crompton, et al. 1979; Duvall, 1990; Maglio, 1978; Romer, 1966; Quiroga, 1979). Yet another branch gave rise to dinosaurs, around

225 million years B.P., who also emerged multi-regionally and appeared world-wide--presumably descending multi-regionally from ancestors who also emerged multi-regionally from multiple ancestral species.

Hence, primordial reptiles split into three lineages, the anapsids which gave rise to modern turtles, synapsids which gave rise to repto-mammals and then therapsids, and diapsids which gave rise to dinosaurs, and birds, and present day reptiles (Caroll, 1988).

Repto-mammals emerged some 250 million years ago, and these creatures briefly ruled the Earth. Although the initial repto-mammals were sprawlers, over time they became physically more refined, and eventually gave rise to therapsids, 200 million years B.P..

Therapsids were exceedingly technologically advanced, physically and neurologically. For example, in contrast to reptiles and amphibians, the elbows were now directed backward and the knees forward which greatly improved their ability to run and manipulate their limbs. In addition, the legs were now located beneath rather than alongside the body which enabled them to run long distances without compressing the chest and lung which allowed them to simultaneously breath while chasing prey. Reptiles must stop in order to breath since their legs, situated alongside their body and chest cavity, constrict the expansion of the lungs as they run.

The therapsids also developed a secondary bony palate which enabled them to chew food and to simultaneously breathe without danger of choking to death. Reptiles must cease to breathe in order to swallow large chunks of their food.

Another advantage occurred in regard to thermoregulation. Therapsids became warm blooded. Reptiles must sun themselves or run around and rely on behavioral thermoregulation. For example, if a reptile fails to move from a cold to a warm location (or vice versa) their body temperature soon approaches that of the external environment. They must move about in order to gain heat by sunning themselves, or cool off by sitting in the shade.

By contrast, the limbic system of the repto-mammals and then the therapsids evolved a means of regulating body temperature internally. Whereas lizards, frogs, fish, etc., have only scales, the therapsids also evolved a coat of fur, as well as sweat glands that release excessive internal heat.

Moreover, with the metamorphosis of the repto-mammal therapsids, the ear underwent important modifications and the limbic forebrain began to expand with new tissues emerging and consisting of additional layers. Because the inner ear and additional brain tissue had emerged, vocalized communication assumed a new importance.

The increased importance of vocalization was made possible by the expanded development of the 4-5 layered cingulate gyrus; a structure that caps the 3-layered allocortical limbic system, and which is implicated in maternal offspring behavior and vocal communication (Joseph, 1999b, 2000a).

With the development of the cingulate gyrus, the therapsid's ability to communicate expanded beyond simple gestures, posturing, or olfactory-pheromonal signaling, and now included the capacity to produce a variety of complex meaningful sounds, such as between mother and infant, including, perhaps the separation cry. The five-layered cingulate gyrus provided the brain power to engage in prolonged maternal care, which in turn promoted the development of language, love, and the family (Joseph, 1993; MacLean, 1990).

Nevertheless, although exceedingly advanced, the repto-mammals were struck down and nearly became extinct (as did almost 95% of all life) following a great cataclysm when the Earth was twice struck by massive meteorites, around 250 million and 225 million years B.P. (Rampino & Haggerty, 1994). These catastrophes were followed by a "giant volcanic eruption"--all of which acted to split apart the already fracture land masses and to blot out the sun with dust, thus dropping temperatures and killing off over 50% of all marine life and all larger size terrestrial animals. These catastrophes killed off most but not all of the larger sized repto-mammals, and gave the much smaller dinosaurs a competitive advantage. The remaining repto-mammals were displaced by these "terrible lizards" who then began to evolve multi-regionally on every continent.

Those dinosaurs who evolved multi-regionally included the 36 foot-long Suchomimus tenernsis, who had teeth shaped like steak knives, and whose fossil remains have been discovered in Egypt, Brazil, England, and central Niger in Africa. Likewise, although Tyrannosaurus rex, the most fearsome meat eater in history, is associated with the Americas, similar species emerged in Asia (Cifelli et al., 1997). According to Cifelli, "nearly all the dinosaurs we find in Utah were either first seen in Asia." In other words, these cold blood raptors evolved multi-regionally (that is from multiple ancestral "seeds") as it not highly likely that they swam the oceans from Asia or crossed over the frozen tundra of the Arctic (being a cold-blooded species) in order to appear in the Americas.

Following every mass extinction, although the majority of species are typically wiped out, others manage to recover. As the environment acts on gene selection, and as the remaining repto-mammals were relegated to a nighttime environment, their brain was forced to further evolve. They adapted to lurking about at night and hiding beneath deep foliage during the day--a lifestyle which induced further expansions in the olfactory dominated forebrain which became increasingly adapted to process olfactory and auditory cues.

The Environment And Therapsid Multi-Regional Metamorphosis

By 200 to 150 million years B.P., the repto-mammals had become therapsids (Caroll, 1988). In addition to the other changes already mentioned, the therapsid brain (the dorsal pallium) was now capped with a five layered (mesocortical) cingulate gyrus--the evolution of which ushered in a revolution in vocal-emotional communication and infant-maternal behavior (Joseph, 1993; MacLean, 1990).

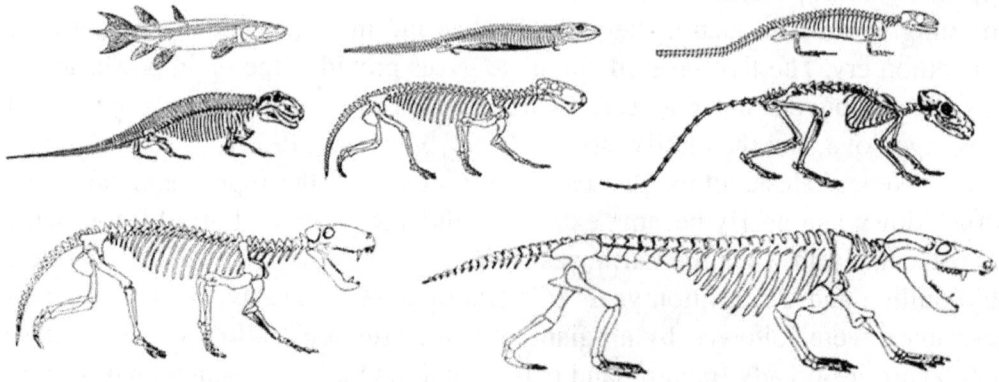

By 150 to 85 million years ago, various suborders of therapsids had given rise to the intellectually and neurologically superior mammals, who in addition to a five layered cingulate gyrus, had evolved a six layered neocortex. Yet, because the environment acts on gene selection, those therapsids living in sheltered pockets of primeval swamp and jungle, remained therapsids--just as n some pockets of the world, Homo erectus remained Homo erectus, and Neanderthals remained Neanderthals, although Cro-Magnon were beginning to swarm over the planet.

Because the environment acts on gene selection, different environmental and climatic conditions can produce not only diverse subspecies, but enhance or slow the rate of species metamorphosis depending on where they dwell. Metamorphosis occurs at different rates under different geological and climatic conditions, and thus at different times periods for the same species, and often not at all. An unvarying environment, coupled with related genetic factors hinders the development of the next stage of metamorphosis.

Hence, in a few isolated swamps and jungles of the world which have undergone little change over the course of the last several hundred million years, huge insects, amphibians and reptiles abound and jungle dwelling mammals still lay eggs -much like repto-mammals.

Egg laying "mammals" (monotremes), tend to be found only in a few isolated regions of the world such as in the primeval swamps and jungles of New Zealand. These egg layers include the anteaters and the even more primitive duck billed platypus. Monotremes are in fact quite primitive, and appear in the fossil record as far back as the earliest periods of the Pleistocene.

By contrast, throughout much of the rest of the world, species of amphibians and reptiles have diminished in number and diversity, and monotremes have disappeared, whereas mammals have climbed the next step of the evolutionary ladder. The egg and the embryo are now nourished inside the womb (placentals).

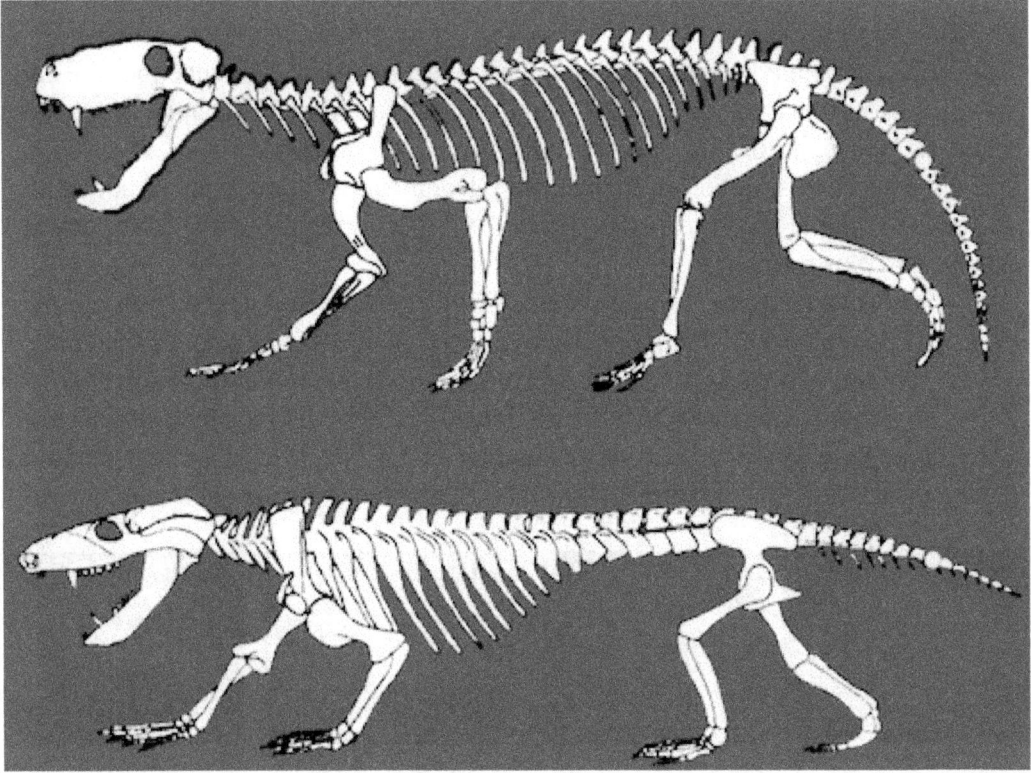

Egg laying mammals, the monotremes, therefore, are more like therapsids and repto-mammals than true mammals (placentals). Like repto-mammals, mono-tremes not only lay eggs, but are without breasts. Instead, they suckle their young via modified sweat glands which secrete milk. Over the course of evolutionary metamorphosis these sweat glands eventually became the mammallary glands of the more advanced mammals (Duvall, 1988).

The monotremes, therefore, are a type of very advanced repto-mammal, a therapsid which has yet to reach the next stage of metamorphosis. In other lands and environments, however, the monotremes have disappeared; replaced or killed off by more advanced mammals who emerged almost simultaneously and multi-regionally throughout Africa, Eurasia, and North America.

These advanced mammals, however, did not simply crawl out of the earth, or emerge from stone or clay. Advanced mammals are the multi-regional descen-dants of therapsids who are the multi-regional descendants of repto-mammals, who in turn, once laid eggs.

Given that repto-mammals, therapsids, and mammals (including primates) have emerged on every continent, it could be argued that the genetic seeds to pro-duce mammals have matured at different rates, albeit in different environments. (Joseph, 1997). Environmental factors coupled with multiple trees of life, explains why modern human mammals still share the planet with primitive egg layers who are little different from their repto-mammal grandparents who strutted their stuff 200 million years ago.

Since the environment acts on gene selection it can influence the rate and speed of evolutionary metamorphosis and the activation and exchange of genetic material. Hence, it not surprising that primitive and modern versions of the same species may coexist, albeit in different environments, e.g. an isolated, steaming swamp and jungle, vs the fruited plains and happy hunting grounds of the Eurasia, Africa, and North America.

Those who appear to be more primitive and who have lagged behind, live in an environment which has not promoted the next stage of genetic metamorphosis. Those "genetic seeds" have yet to mature. In fact, this same unequal relationship where primitive and more advanced species coexist, albeit in different environments, is characteristic of all manner of Earthly life, including bacteria, plants, insects, reptiles, mammals, primates, and even the genus Homo.

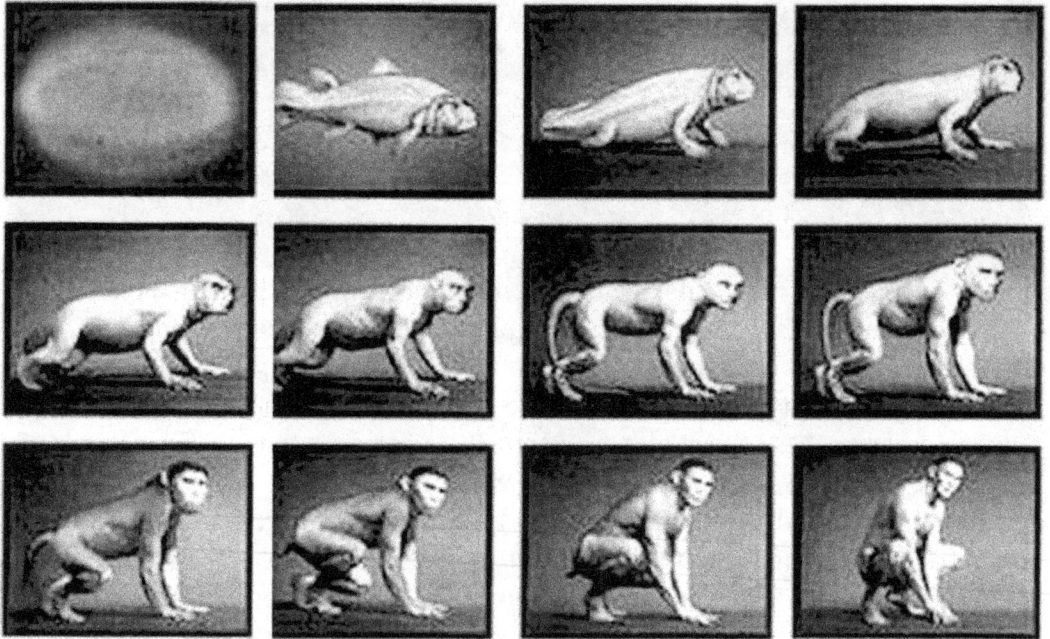

Mammalian Metamorphosis

As noted, with the development of the cingulate gyrus, the therapsid's ability to communicate expanded beyond simple gestures, posturing, or olfactory-pheromonal signaling, and now included the capacity to produce a variety of complex meaningful sounds, such as between mother and infant, including, perhaps the separation cry. The five-layered cingulate gyrus provided the brain power to engage in prolonged maternal care, which in turn promoted the development of language, love, and the family (Joseph, 1993; MacLean, 1990).

As therapsids (e.g., Probainognathus from the Triassic followed by Periptychus from the Paleocene), continued to evolve and to become more intelligent, the mesocortical five-layered cingulate began to sprout a small nub of neocortex (Quiroga, 1980); i.e. the six layered new cortex. In later appearing therapsids, e.g. Phenocodus, the now, enlarged brain, began to resemble that of primitive mammals, e.g., opossum or hedgehog.

When the first mammals began to scurry about, over 85-130 million years ago (e.g. Kumar & Hedges, 1998), the gray mantle of the outer surface of the brain, the six layered neocortex ("new cortex") had begun to encapsulate the old brain, forming the frontal, parietal, temporal, and occipital lobes in the process.

It was the development of this new brain and the neocortex which provided mammals and primates with a enormous competitive intellectual edge that enabled them to take advantage of the cosmic catastrophe which presumably wiped out most of the large, cold blood, land based dinosaurs when a massive meteor struck the Yucatan peninsula, some 65 million years ago (Alvarez, 1986; Alvarez & Asaro, 1990; Rampino & Haggerty, 1994; Raup, 1991).

The enormous energy released from this meteor strike destroyed much of life in the Americas. Moreover, due to the dust thrown into the air sunlight was blocked out for months. Temperatures dropped, thus killing off all remaining large sized cold blood animals; events which were then followed by an acid rain and a greenhouse type warming.

Any remaining dinosaurs were quickly eradicated by surviving mammals, and in consequence, mammals gained dominion over the day as well as the night. As the environment acts on gene selection, the mammalian brain quickly adapted to processing visual as well as auditory stimuli, and expanded yet again.

With the ensuing evolution of primates, monkeys and apes in particular, the entire forebrain became adapted for engaging in prolonged and detailed analysis of visual and auditory stimuli (Gloor, 1997; Stephan, 1983) as well as climbing in trees.

Over the course of later mammalian and primate evolution and as these creatures gained complete dominion over much of the planet, the neocortex began to expand at a rapid rate (Stephan & Andy, 1977). In fact, when comparing the brains of "living fossils" such as insectivores with that of humans, it appears that the six to seven layered neocortex expanded by a factor of 156 (even when taking into account differences in body size), whereas the 3-layered limbic system allocortex and five-layered mesocortex (cingulate gyrus) and all associated olfactory-limbic structures developed at a much reduced rate.

For example, limbic system structures such as hippocampus and septum are only 4 times larger and the amygdala is 6 times larger when comparing humans to insectivores. By contrast, the olfactory bulb is 40 times smaller (Stephan & Andy, 1969; Stephan, 1983), which is due to the reduced importance of smell and olfaction in human behavior.

Hence, contrary to Darwin's theory, the brain did not become more variable, but "evolved" new structures and tissues which mushroomed in size.

Multi-Regional Primate Metamorphosis

It is only with the demise of the dinosaurs that primates were able to emerge from the underbrush and the darkness of night. By 55 million years ago, during the early Eocene, at least some orders of primates (e.g. Tetonius) evolved a large occipital lobe (visual cortex), as well as an emerging temporal (auditory) lobe and frontal lobe (Radinksy, 1967, 1970). Thus, by 55 million years ago, ancestral primates had evolved a brain which resembled that of modern day prosimians primates.

Over the ensuing years, and as primates adapted to living in the trees, which in turn required major adaptations in the eyes and hands, the basic pattern for the primate neocortex became established and the frontal and temporal lobes continued to expand.

The first prosimian primates to scurry about this planet may have diverged from several different mammalian lines some 70-100 million years ago. Primates, like earlier mammals, reptiles, and amphibians, came to live on every continent. Although modern day neo-Darwinian theory demands a single line of descent, like the mammals, dinosaurs, reptiles, amphibians, plants and insects before them, primates appear to have emerged multi-regionally and almost simultaneously throughout the world and thus from multiple branches from multiple trees, rather than from a single seed and a single trunk.

Of course, it is possible that many species may have simply migrated from one land to another, for example, from North American to South America. However, migration becomes less likely as to the emergence of, for example, primates in South America and Africa-Eurasia, as the great oceans that separate these lands is simply to vast a distance to be covered without first dying of thirst and hunger.

Throughout the world, many species of primate took to living amongst the branches of the trees. In consequence tremendous alterations occurred in the fingers, and hand-eye coordination--as the environment acts on gene selection. For example, claws became grasping fingers whereas the eyes moved to the front of the face thus providing for depth perception and stereoscopic vision. Within the brain there were tremendous expansions of the visual, auditory, tactual-gestural neocortex.

It was presumably from these widely dispersed tree loving stocks that gave rise to "old world" monkeys in Africa, India, Asia, and "new world monkeys" in the Americas about 40 million years ago (Leaky, 1976, 1988; Pfeiffer, 1985). The wide ranging stock of "old world" monkey, in turn gave rise to apes (hominoids) about 30 million years B.P., with what would become chimpanzees and gorillas eventually appearing in Africa, and Orangutans appearing in India and Asia.

Presumably numerous branches from these varied primate-hominoid trunk lines diverged again, and yet again, and gave rise to the ancestral lines which led to the emergence of the first primitive Adam and Eve.

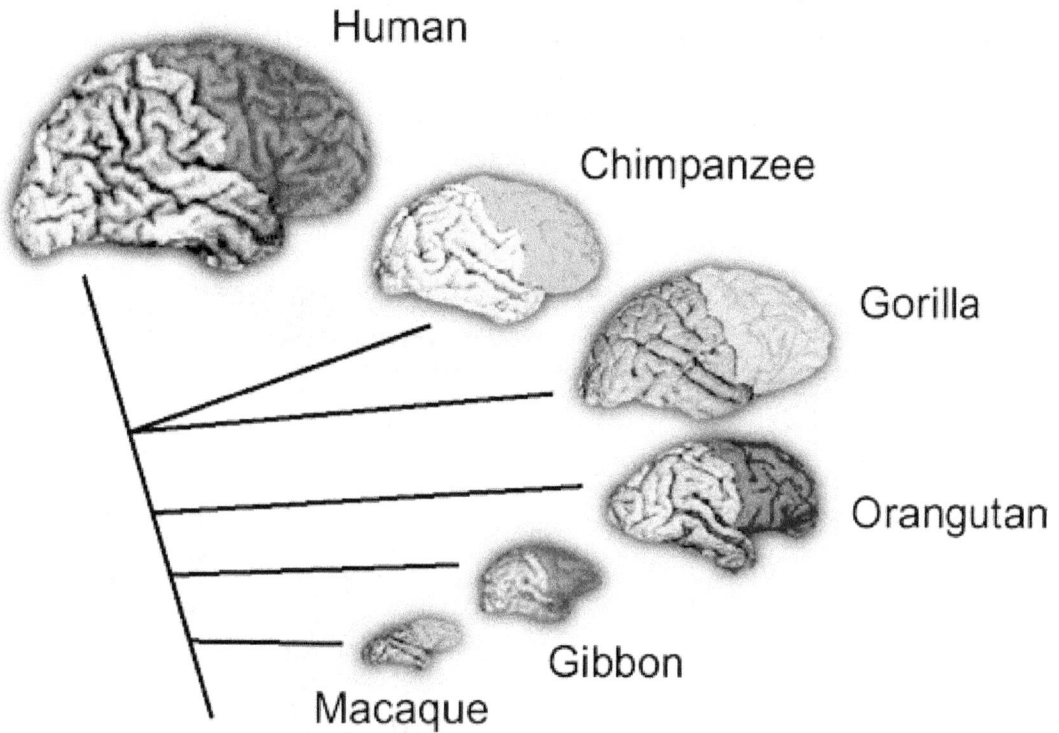

From Hominoid To Hominid

By 30 million years B.P., apes had emerged multi-regionally dwelling in Africa, China, and India, literally from sea to shining sea. And it is from these wide ranging hominoid stocks, that a variety of hominoid/pre-hominids began to "evolve," including Dryopithecus, Sivapithecus, Ramapithecus, Ankarapithecus, Ouranopithecus, and Giganotopithecus.

Evolutionary metamorphosis is most likely to occur when an organism is exposed to a multiplicity of changing environments, or where two divergent worlds meet. For the pre-hominid hominoids such as Dryopithecus, Sivapithecus, Ramapithecus, Ankarapithecus, Ouranopithecus, and Giganotopithecus, the netherworld of change was found where the forest ended and began to recede and the savanna and grasslands began to expand. This great change occurred during a period in which parts of the planet were bathed in renewed warmth.

As the changing environment acts on gene selection, the descendants of these pre-hominid hominoids gave rise to a variety of pre-human species which began to "evolve" multi-regionally. These included Australopithecus who was later followed and joined by Homo habilis who was joined and then followed by H. erectus who was joined and then followed by Homo sapiens who was joined and then followed by H. sapiens sapiens--the wise man who knows he is wise and who would soon dominate and then threaten a good part of the planet's multiple life forms with death and extinction.

Multi-Regional Human Metamorphosis: From Hominoid To Hominid

As to the ancestors of the first pre-human hominids, there are several candidates, each of which may have given rise to a distinct or similar branch of the emerging human race. These ancestral species include Dryopithecus, Sivapithecus--ape-like hominid/hominoids who emerged in Europe and India, about 16 million years ago. Other candidates include Ramapithecus whose remains have been discovered in Africa, India, and Southwest China (Jurmain, et al. 1990; Munthe et al. 1983). Ramapithecus, in fact, appears closely related to Dryopithecus and Sivapithecus.

Other possible candidates include Ankarapithecus of Turkey, Ourano-pithecus of Greece, and Giganotopithecus whose 8 million year old remains have been found in India, China, and Vietnam (reviewed in Howell, 1997). Giganotop-ithecus may have descended from Ramapithecus and may have later given rise to Homo habilis in Asia.

Nevertheless, these species of hominoid pre-hominid have for the most part been rejected as human ancestors. Conventional wisdom requires an African origin for the proverbial ape-hominid-human ancestor, and these species lived in the "wrong" parts of the world. Indeed, conventional wisdom requires that the facts fit the theory, and those facts and fossil remains which are inconvenient or inconsistent with accepted theory, such as the evidence for multi-regional meta-morphosis, are conveniently rejected and dismissed.

Multi-Regional Metamorphosis Of Australopithecus & Homo Habilis

Around five million years ago and in reaction to yet another major change in environmental and climatic conditions, some species of hominoids began to increasingly live upon the ground. Although they spent much of their time in trees, it is while on the ground, around 5 million years ago, that the descendants of Ramapithecus and/or Giganotopithecus Ankarapithecus, Ouranopithecus, or some other primate-pre-hominid, underwent further evolutionary metamorphosis and gave rise to a variety of more advanced pre-hominids, such as Ardipithecus ramidus and Australopithecus Afarensis.

Again, however, contrary to conventional wisdom, but consistent with the theory of evolutionary metamorphosis, a wide range of species collectively re-ferred to as Australopithecus (A. aethopicus, A. africanus, A. robustus, A. boisei), emerged multi-regionally, throughout Africa (see Grine, 1988; Leakey & Walker, 1988; Skelton & McHenry 1992) as well as in China and Java, e.g. A. robustus (Barnes, 1993).

Around 2-3 million years ago Australopithecus was joined by other pos-sible human ancestors: Homo habilis (the handy man). Again, contrary to con-ventional wisdom, but consistent with the theory of evolutionary metamorphosis, several varieties of H. habilis (e.g. H. rudolfensis, H ergaster) appeared in Africa, as well as in China (Dragon Hill) and Indonesia (reviewed in Barnes, 1993, and Howells, 1997). Like their purported ancestors, these species collectively referred to as Australopithecus and H. habilis "evolved" multi-regionally . However, un-like their predecessors they evolved the capacity to stand upright and to make and use tools.

The Multi-Regional Metamorphosis Of Homo Erectus

Like their ancestors, the fossil evidence indicates that Australopithecus and Homo habilis continued to evolve multi-regionally and to undergo metamor-phosis in various parts of the world, thus giving rise to a wide variety and wide ranging species collectively referred to as Homo erectus (Binford, 1981; Brown, et al. 1985; Jia, 1980; Johanson & Shreeve, 1989; Leaky, 1976, 1982; Pfeiffer, 1985; Rightmire, 1990; Swisher et al., 1996; Stanley 1979, 1981).

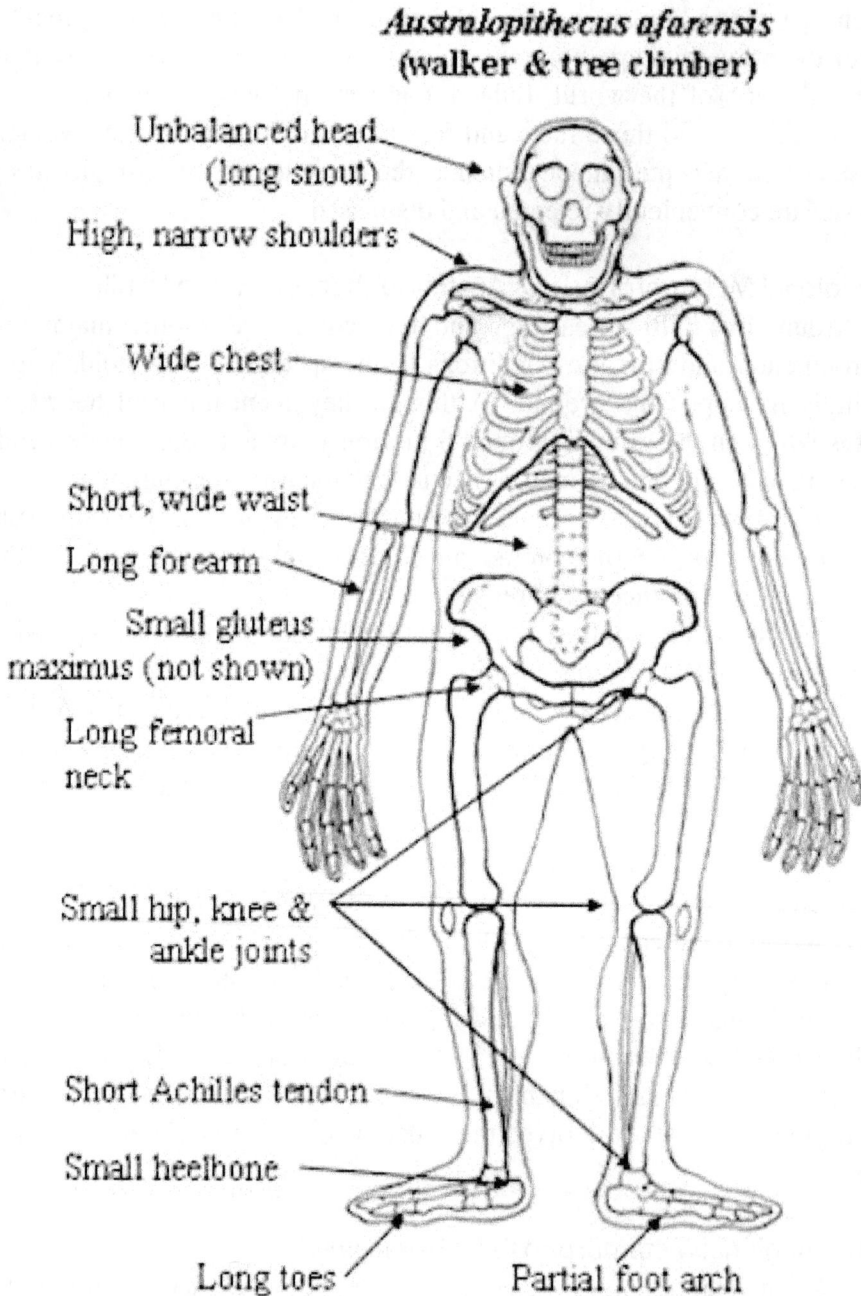

Australopithecus afarensis
(walker & tree climber)

Unbalanced head
(long snout)

High, narrow shoulders

Wide chest

Short, wide waist

Long forearm

Small gluteus
maximus (not shown)

Long femoral
neck

Small hip, knee &
ankle joints

Short Achilles tendon

Small heelbone

Long toes

Partial foot arch

Homo erectus
(walker & endurance runner)

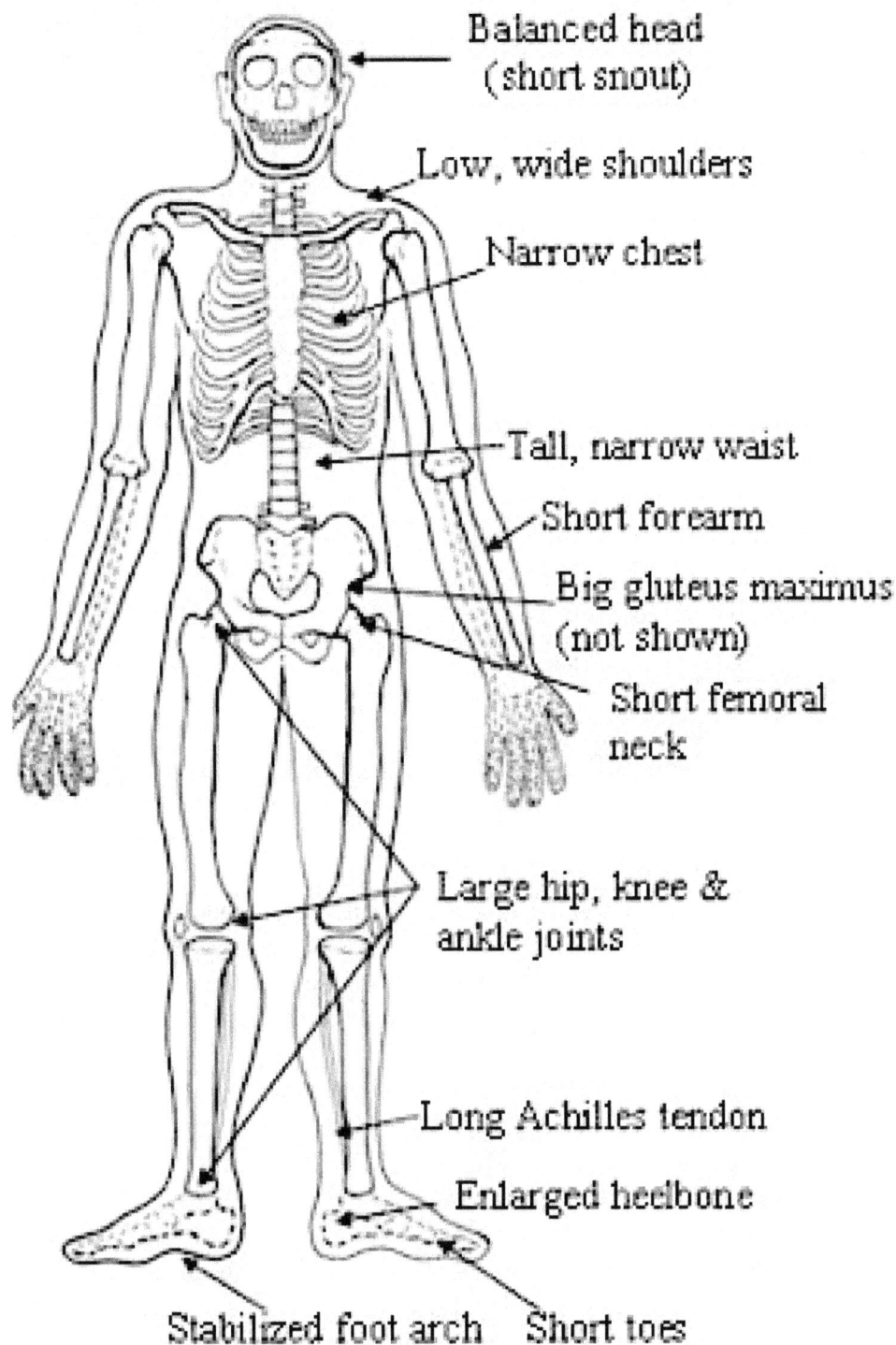

Balanced head
(short snout)

Low, wide shoulders

Narrow chest

Tall, narrow waist

Short forearm

Big gluteus maximus
(not shown)

Short femoral
neck

Large hip, knee &
ankle joints

Long Achilles tendon

Enlarged heelbone

Stabilized foot arch   Short toes

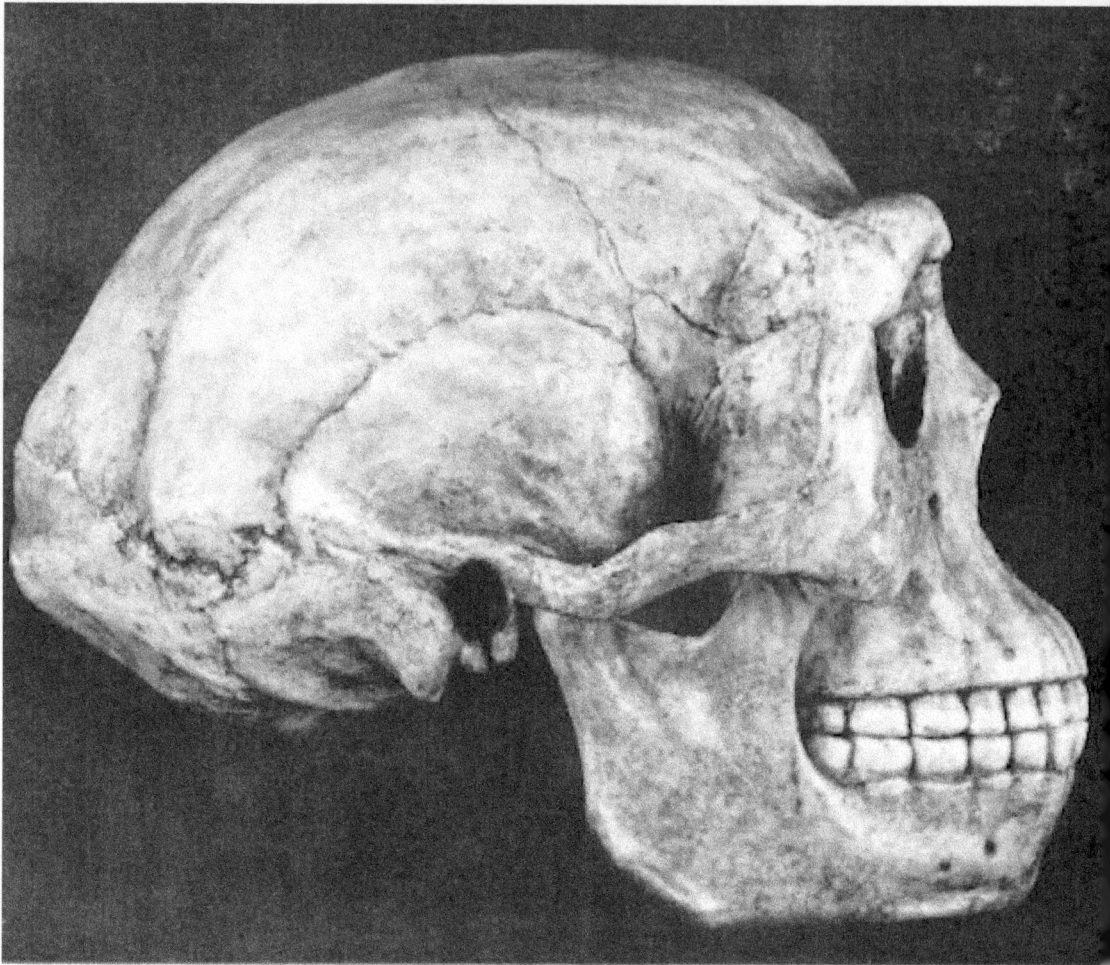

With the metamorphosis of Homo erectus, who first appeared around 1.9 million years ago, the brain gradually doubled in size, approaching within 15% of a modern human by 500,000 years B.P.

The Homo erectus were the first individuals to have harnessed fire and the first who developed crude shelters and home bases. They also utilized various earth pigments (ochre) for perhaps cosmetic or artistic purposes. In this regard, these peoples were beginning to experiment with individual creative and artistic expression.

However, contrary to conventional wisdom which requires an African origin, the earliest fossil evidence indicates that Homo erectus emerged multi-regionally, beginning in Euro-Asia, i.e., the Caucasus (Georgian Republic) around 1.9 million years (reviewed in Howells, 1997).

H. erectus emerged multi-regionally in Asia, Africa and Europe. The skeletal remains of H. erectus (and associated stone tools) have been discovered in Ceprano Italy (dated to 800,000 B.P.), and in Java, Indonesia, (Pithecanthropus erectus) and near the Solo River (e.g. Solo Man, Java Man)--sites dated from 1.8 million to 700,000 B.P (respectively) Likewise, H. erectus (H. erectus pekinesis) have been discovered in Northern China (e.g. Peking Man) and in Zhoukoudian, Yuanmou and Xihouda China--sites dated from 500,000 to 750,000 to 800,000 to 1.5 million years B.P. (Jia, 1975, 1980; Jurmain et al. 1990; Stanley, 1979, 1981; Wu & Wang, 1985).

Again, although conventional wisdom and neo-Darwinian theory requires it, it is not likely that "Georgia Man," "Solo/Java Man" and "Peking Man" migrated out of Africa. The fossil remains dated to 1.6 to 1.8 million years in Java, and 1.5 to 2 million years in China (Dragon Hill), and the jaw from a H. erectus discovered in the Caucasus dated to 1.9 million years, are in fact older than similar specimens found in Africa. In fact, consistent with the theory of multi-regional metamorphosis, the species of Homo erectus discovered in China may well trace its lineage to Gigantopithecus who may have given rise to an Asian H. habilis (Jia, 1980; Munthe et al. 1983) who in turn gave rise to an Asian Homo erectus, as the remains of these ancestral species have also been discovered in Asia.

Hence, contrary to the out-of-Africa scenario, it could be argued that H. erectus first emerged in the Caucasus or Java, or Asia, or all of the above, and either spread to Africa (where H. habilis was still the dominant hominid), and finally Europe, or independently and multi-regionally evolved from ancestral species living in these other lands.

The Multi-Regional Metamorphosis Of Archaic Homo Sapiens

Just as H. erectus appears to have evolved multi-regionally, the fossil evidence indicates that by 350,000 years ago, Archaic humans had emerged multi-regionally, in Africa, Asia, Europe, the Middle East, and India. In Europe archaic remains have been discovered in Petralona Italy, from sites dated between 350,000 to 400,000 B.P., as well as in Hungary and Germany. Likewise, the remains of archaic H. sapiens have been discovered in China (Hupei Province) dated to at least 350,000 B.P.

Moreover, multi-regional progressive evolution is evident as more advanced and modern appearing archaics ("early moderns") have been discovered in China (Dingcun, Maba, Dali, Jinniushan) as well as in East Africa from sites dated to 130,000 B.P. and 120,000 B.P respectively (Barnes, 1993; Butzer, 1982; Grun et al. 1990; Howells, 1997; Rightmire, 1984). Again, however, the more advanced species appeared in China first.

Hence, there is a clear line of multi-regional descent. Australopithecus, H. habilis, H. erectus, archaic and early modern H. sapiens appear to have emerged multi-regionally, with three distinct species of humanity, erectus, archaic, and "early moderns" sharing different regions of the planet simultaneously.

What the evidence and fossil record indicates is that the Earth was genetically seeded to grow humans (and other species) and all manner of variations thereof. Rather than a single tree with a single trunk, the "seeds" for a forest of similar trees rained down upon the planet and over times gave rise to similar, but not identical species. As the environment acts on gene selection, these "trees" have eventually bore a variety of similar (but not identical) "fruit" albeit at different times and at different rates, depending on environmental conditions.

The Multi-Regional Metamorphosis Of Modern Humans

The environment acts on gene selection, genes which exist prior to their activation. Differing environments, therefore, can induce slight changes in species living even in adjacent lands, or alterations so significant, that a wholly new species emerges and displaces all locally situated ancestral species. Moreover, pockets of basically the same species may emerge multi-regionally, albeit some at an earlier or later date than others as is evident in regard to Homo erectus, Neanderthals, and the Cro-Magnons peoples, all of whom shared the planet during overlapping time periods.

Hence, whereas archaic Homo sapiens first emerged almost 500,000 years ago, large populations of Homo erectus continued to dominate parts of the planet until 300,000 years B.P., with a few isolated populations possibly hanging on until just 27,000 years ago as discovered in the island of Java (Swisher et al., 1996). Thus, for almost 200,000 years, large populations of Homo erectus and the more advanced archaic Homo sapiens shared the planet, albeit in different geological locations.

Moreover, just as Australopithecus, H. habilis, and H. erectus appear to have evolved multi-regionally, the fossil evidence suggests the same for archaic H. sapiens, whose remains have been discovered in Africa, Asia, Europe, the Middle East, and India. And, contrary to the single seed, out-of-Africa scenario, the remains of evolutionarily advanced archaics ("early moderns") appear in China (Dingcun, Maba, Dali, Jinniushan) 10,000 years before similar species appear in East Africa; i.e., from sites dated to 130,000 B.P. and 120,000 B.P respectively (Barnes, 1993; Butzer, 1982; Grun et al. 1990; Howells, 1997; Rightmire, 1984).

Archaic H. sapiens in fact died out before the more primitive H. erectus, i.e. 29,000 B.P., (Neanderthals) vs 27,000 B.P. (H. erectus), whereas another species of incredibly advanced humans, Homo sapiens sapiens, had already emerged 75,000 years B.P. And, the first H. sapiens sapiens did not first appear in Africa, but in Australia (75,000, B.P.) followed by China (67,000 B.P.), Israel, Romania, and Bulgaria (43,000 B.P.), Iraq and Siberia (40,000 B.P.), Spain and France (35,000 B.P.), and Brazil, Peru, Chile and North America (30,000 to 50,000 B.P.). By contrast, during these same time periods, sub-Sahara Africa was still populated with archaic Homo sapiens; African Neanderthals.

Indeed, rather than originating in Africa where archaic H. sapiens roamed until 30,000 B.P., "modern" H. sapiens sapiens had already established numerous settlements in Australia and were fashioning complex tools as early as 60,000 B. P., including grooved "waisted blades" which could be bound to a handle.

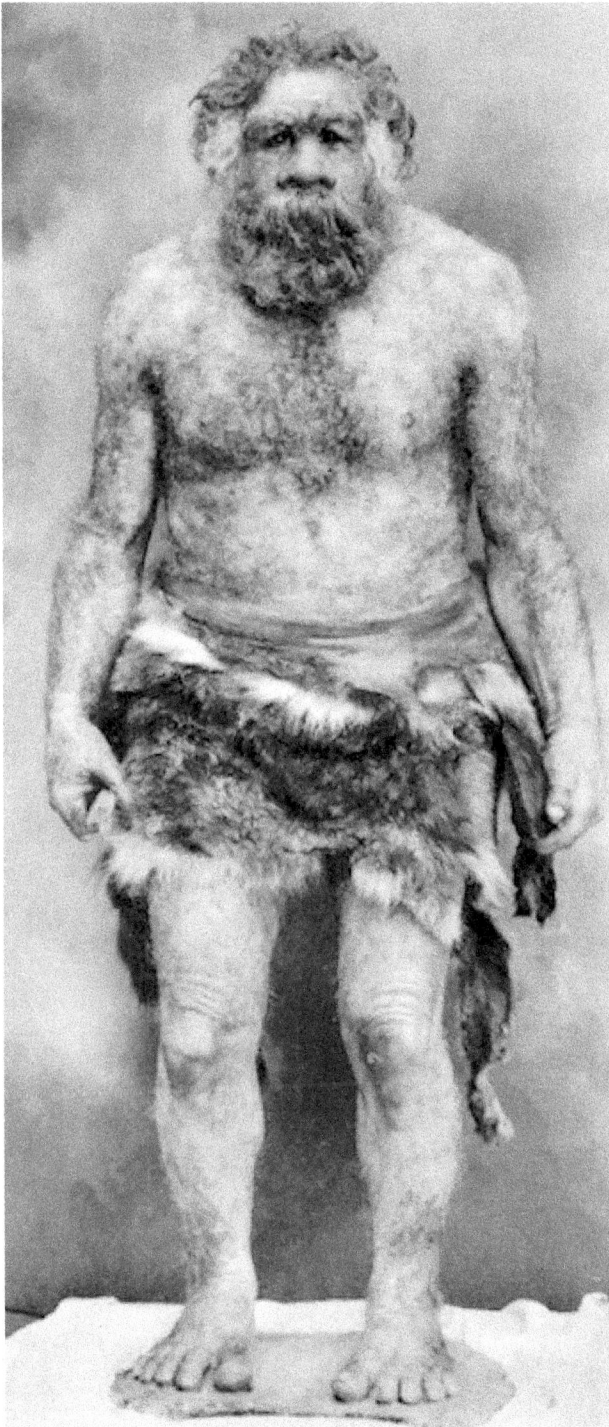

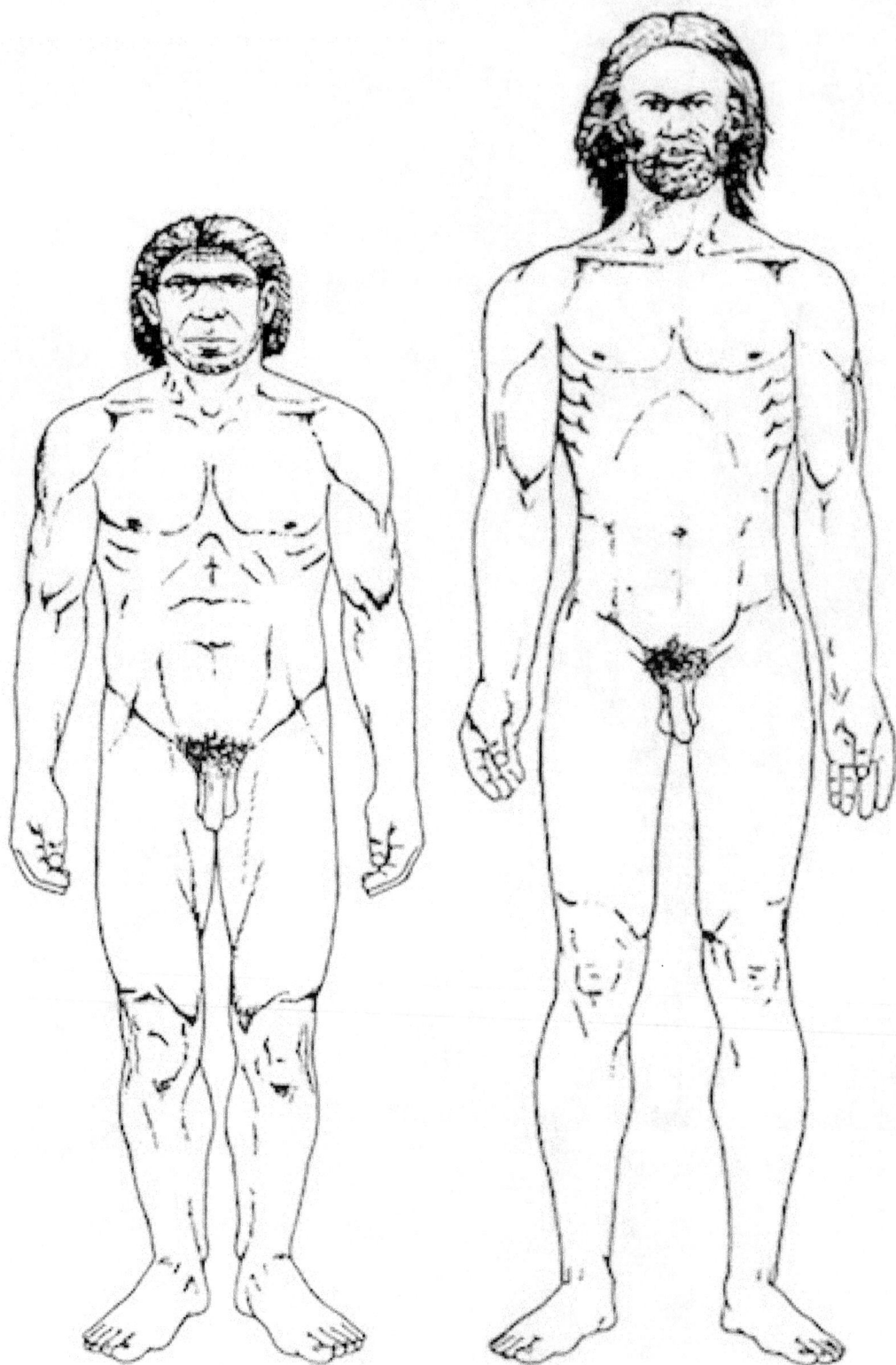

Alien Viruses, Genetics, Cambrian Explosion: Humans

What this means is that four different species of humanity were living in different parts of the world simultaneously, H. erectus, archaic, early modern, and "modern" H. sapiens sapiens--which is evidence of multi-regional evolution occurring at different rates in different branches of humanity, in different geographical regions. Although it could be argued that the descendants of one of these geographical groups merely migrated and killed off all competitors, this view is not supported by the fact that different types of humanity, for example, different types of H. habilis and H. erectus, are found in Africa vs Asia, with those in Asia (e.g. H. erectus) appearing before those in Africa, or those in Australia (e.g. "moderns") appearing before those in Africa or Asia.

Australian as well as Asian "moderns" in fact appeared tens of thousands of years before their counterparts in Africa; findings which are not consistent with the out-of Africa scenario but which instead supports the multi-regional view and even the out-of-Asia or out-of-Australia view of evolution. In fact, whereas "modern" appearing H. sapiens sapiens do not emerge in North East Africa until around 35,000 B.P. the skeletal remains of Asian moderns have been found in southern China from sites dated as long ago as 67,000 B.P. (reviewed in Howell, 1997).

In fact, not only do "modern" humans appear outside of Africa thousands of years before "moderns" appear in Africa, but **evolutionarily advanced humans were living in Northern Siberia as long ago as 250,000 to 300,000 years B.P. (Waters, 1997). Siberia is an exceedingly hostile environment requiring advanced survival skills as temperatures drop to below 70 degrees in winter. In fact, stone tools dated to 250,000 years B.P., were discovered along a river near Irutsk, Siberia--tools similar to those found in North America.

"What this indicates," according to Michael Waters of Texas A. & M., who helped date one of these sites and associated artifacts, "is that these people had the ability to deal with a rigorous environment. They could control fire, they had a survival strategy, they could make and find shelter, clothing, boots, etc." However, these advanced behaviors and capabilities, such as the ability to fashion complex clothing, do not appear in sub-Sahara Africa until near the end of the Upper Paleolithic.

Hence, similar to the step-wise world-wide pattern of multi-regional, multi-phylectic metamorphosis which has characterized the progressive emergence and increased complexity of plants and animals (Joseph, 1997, 2000a), the available evidence suggests that human "evolution" has unfolded multi-regionally in a step wise, progressive fashion, with some groups lagging far behind and others being left behind altogether and becoming extinct.

The Earth (and other planets) were genetically seeded to grow humans, and all manner of variations thereof.

The Myth Of African Origins: The Story Of Eve

Conventional wisdom is of a single line of descent, and that modern humans "evolved" in Africa, from earlier species of humanity who also "evolved" in Africa, with each successive species migrating out of Africa, and then killing off and replacing earlier species who had also originated and migrated out of Africa before them.

Although one of the central tenants of Darwin's theory is that of "random variation" it has yet to be explained why that variation only occurred in Africa, and why that variation did not lead just to variable species, but increasingly intelligent and resourceful, and more advanced species. Variation does not equal increased complexity and the fossil record is indicative of a step-wise sometimes leaping progression.

Harvard paleontologist, Gould, solves this problem by denying the obvious and by claiming there is no evidence for progress. Yet others refer to mutations. Random mutations resulted in the production of successive superior species simply by chance, and these mutations repeatedly took place only in Africa.

Thus, according to modern neo-Darwinian theory, mutations were being continually produced in Africa and only in Africa, and each subsequent race of mutants migrated out of Africa only to be later replaced by more advanced mutants who also mutated in Africa.

Not only is the out-of-Africa, single seed, mutation scenario contradicted by the fossil evidence, but by logic and genetics.

For example, a "mutated" gene is generally eliminated before it has a chance to be expressed. It would be eliminated and replaced by a normal duplicate gene.

Moreover, even if the mutant gene were not eliminated, in order to produce a viable breeding pair, requires that two of the same exact mutations appear simultaneously in the same population, in both a man and a woman--and by chance--otherwise, the mutated individual might be unable to breed and pass on his/her superior mutated traits. Nevertheless, even if this mutated individual did breed, the mutation would likely disappear from the gene pool; either that or perhaps only an intermediate individual would be produced.

Indeed, even if we disregard those genetic mechanisms which eliminate "mutations" the fact remains that for evolution to be successful requires that multiple individuals and thus multiple mating partners "step-forward" as a group to the next stage of species evolution in order to produce viable offspring. A scenario such as this, however, is not consistent with neoDarwinian theory, though it is entirely compatible with the theory of evolutionary metamorphosis. We will return to the mutation myth of evolution in the next chapter.

The Trouble With Eve

According to the "Eve" hypothesis (Stoneking & Cann, 1989; Vigilant et al. 1991), all modern humans descended from female ancestors living in Africa about 250,000 years ago--which, coincidentally is at about the same time that the ancient Sumerians claim the Anunnaki first "lowered the god-head" from the heavens.

There are numerous problems with the "Eve" theory, beginning with the fact that the assumptions upon which it rests have been shown to be invalid; e.g., that mitochondrial DNA (which in humans consists of 37 genes) is only inherited from the mother. In fact, fathers also contribute mitochondrial DNA. In addition, the estimates based on mitochondrial DNA "mutation" rates has been shown to be statistically erroneous (Templeton, 1992; Wolpoff & Thorne, 1991).

Specifically, based on an initial analysis of a small fragment (610 base pairs) of the mitochondrial genome (which consists of about 16500 base pairs), taken from 189 individuals, it was argued that there is greater diversity within Africa than outside Africa--as based on the varying patterns in sequencing (Stoneking & Cann, 1989; Vigilant et al. 1991). From this data it was concluded that all humans must have descended from African ancestors. However, others have found, using the same data, that all humans could have also descended from ancestors who lived in New Guinea (Ruvolo & Swafford, 1993).

As per more recent data provided by Chu et al., (1998), regarding the origins of modern Chinese, it is noteworthy that although these investigators also claim an African origin, that the East Asian population they studied were genetically more closely related to "Native American" Indians, followed by Australian aborigines, and New Guineans. Hence, this data could also be interpreted to mean that anatomically "modern" humans originated in Australia and then migrated to New Guinea, then to southern Asia, and then the Americas.

And then there is the Neanderthal problem. Neanderthals did not evolve into Cro-Magnon peoples, as they coexisted for at least 20,000 or more years. And, European Neanderthals are genetically unrelated to modern Europeans (Krings et al., 1997; Ovchinnikov et al., 2000). If all species of humanity first evolved and then migrated out of Africa, how is it that two separate samples of DNA from different Neanderthals living in distant lands indicate that although these peoples were closely related, they are unrelated to modern peoples living in Europe? The answer? Because they evolved from an ancestral "tree" that significantly differed from that of the Cro-Magnon.

Again, the environment acts on gene selection, and slight changes in the environment can activate different genes and gene sequences, thus producing variant versions of the same species. However, variant versions of the same species can only be produced if both species possessed the same genes that coded for the same species--or variations thereof.

The evidence based on Neanderthal DNA, like the fossil record, is consistent with the theory of evolutionary metamorphosis. The Earth (and other planets) were genetically seeded to grow humans, and all manner of variations thereof.

And why didn't the Neanderthals evolve into modern people? First and foremost, they dwelled in Europe during an epoch of extreme Arctic cold--a bleak and frigid environment which limited experiential opportunities. They also failed to evolve because once the weather began to change, the Cro-Magnons moved in, and the Neanderthals were either eradicated by the intellectually and technologically superior Cro-Magnons, and/or they died out due to the diseases that the Cro-Magnons brought with them as they invaded Neanderthal lands.

These later possibilities are also consistent with the theory of multi-regional evolutionary metamorphosis--as is the Sumerian claim that "gods" from other planets genetically altered some of the primitive humans living upon the Earth (e.g., Neanderthals), in order to create an intellectually superior being fashioned in the image of the gods, but who could serve as slave labor, i.e., the Cro-Magnon. According to the Sumerians, the men and women created in the image of the gods were also exceedingly sexually prolific, and their population mushroomed out of control. In contrast, according to the Sumerians, the more primitive species of humanity were sexually exceedingly primitive (suggesting that Neanderthal women had not yet lost their estrus) and once they came into contact with the god-like human creations of the gods, the Neanderthals died out as a species.

Evolutionary Metamorphosis

Only the purposeful expression of genetically coded instructions can account for the obvious evidence of a step-wise, sometimes leaping progression in increasing intelligence and complexity which has characterized the metamorphosis of a rather narrow range of life on this planet. Only precise genetic instructions can account for the fact that basically similar species have emerged multi-regionally across distant lands, from distinct pockets of ancestral species which also emerged multi-regionally, and this includes the multi-regional metamorphosis of human beings. The planet (and others like it) was genetically seeded to grow all manner of species, including humans; and these genetic instructions were maintained in the genomes of the first creatures to be flung upon the face of this planet billions of years ago.

Over the course of "evolution" and the genetic engineering of the earthly environment, the unlocking and release of these "genetic memories" and silent traits, has resulted in the multi-regional replication of creatures (or variations thereof) who may well have lived on other planets, including fish... frogs... reptiles... repto-mammals... mammals... primates... and finally woman, man....

Human ancestry leads interminably into the long ago.

9. Evolution in the Ancient Corners of the Cosmos: Even the Gods Have Gods Who Have Gods

The Complexity of Life: Life Could Not Have Begun On Earth

Life was present on this planet from the very beginning, with evidence of biological activity present, dated from 4.28 bya to 3.8 bya during a period when the surface of Earth was being repeatedly pulverized and vaporized by a constant bombardment of meteors, comets, and cosmic debris. Given this Earth is approximately 4.6 billion years old, the lack of essential ingredients for creating RNA, DNA and basic proteins, the thermal conditions which would have destroyed all biological molecules, coupled with evidence of life within 300 to 800 million years after earth became Earth, the likelihood that life could have been fashioned on this planet during this time is beyond improbable.

Single cellular microbes are comprised of more than 2,500 small molecules (e.g. including amino acids consisting of 10 to 50 tightly packed atoms), as well as macro-molecules (proteins and nucleic acids) and polymeric molecules (which are comprised of hundreds to thousands of small molecules) all of which are precisely jigsawed together to form a single celled organism (Cowan & Talaro, 2008). The tiniest and most primitive of single celled creatures contain a variety of micro- macro- and polymeric molecules and over 700 proteins which fit and function together as a living mosaic of tissues. Moreover, each of the many thousands of different molecules that make up a single cellular creature perform an incredible variety of chemical reactions--usually in concert with that cell's other molecules and their protein (enzyme) products.

How could chance combinations have created such complexity, a living mosaic within 300 to 800 million years after the Earth began to form when the lack of essential ingredients and the hell that was Earth would have madeit an impossibility? Nobel laureate Francis Crick (1981) believed that even 10 billion years would not be enough time. Estimates of the time needed for these chance combinations to have produced life have ranged from 100 billion to over 1 trillion years (Crick 1981; Horgan, 1991; Hoyle, 1974, 1982; Yockey 1977) to completely improbable (Dose, 1988; Kuppers, 1990).

The genome of *Mycoplasma genitalium*, one of the smallest free-living microbe (Fraser, et al., 1995), has over 580,000 base pairs and over 213 genes, 182 of these coding for proteins.

Even the simplest of single celled "organisms," such as *Carsonella*, requires 160,000 base-pairs of DNA, and 182 separate genes, in order to live and function

(Nakabachi et al., 2006). However, *Carsonella* cannot live independently, but is parasitic and depends on a living host, a psyllid insect, to survive.

Carsonella may not even be a living entity, but rather an organelle that escaped from or was inserted by a parasitic bacteria (Tamames, et al., 2007). However, even if we classify *Carsonella* as a proto-bacteria or proto-organism, and if it or something similar was created on this planet, randomly by chance combinations abiogenetically in an organic soup or deep sea thermal vent, then this soup had to randomly create, assemble, organize, and then spew out over 182 genes, comprised of over 160,000 base pairs. This is the equivalent of discovering over 180 computers on Jupiter and claiming they were magically assembled in the Methane sea when elementary particles were randomly jumbled together. However, even with 182 genes, the resulting creation could not have survived unless provided with a living host.

Hoyle (1974) calculated that the probability of forming just a single protein consisting of a chain of 300 amino acids is $(1/20)^{300}$ or 1 chance in 2.04×10^{390}. Yockey (1977) calculated that the probability of achieving the linear structure of one protein, 104 amino acids long, by chance is 2×10^{-65}. The odds of this happening on Earth within three hundred millions years, or even within 10 billion years is completely improbable.

A living cell of course, contains more than a single protein.

Microbes range in size, but the smallest free ranging microbes consist of at least 700 proteins (Cowan and Talaro, 2008). However, even if we were to propose that only 240 to 250 proteins were necessary to create the first replicon, or proto-organism, the probability of forming these proteins from left-handed amino acids would be between 1 in $10^{29,345}$ to 1 in $10^{33,635}$. In other words, it would take trillions of chance combinations of all the necessary ingredients, with all the ingredients freely available and concentrated in the same location where the mixing was taking place and under ideal life promoting conditions. And yet, the most crucial elements, such as oxygen, sugar, and phosphorus, were not freely available on the Hadean Earth (Russell & Arndt 2005; Sun, 1982).

Even if the necessary chemicals were available, as a matter of basic statistics, the probability that a single protein, or a single gene, or that life would randomly form on Earth within 300 million or even a billion years, given these odds, is essentially zero.

Specifically, and in accordance with what is known as "Borel's Law" any odds beyond 1 in 10^{50} have a zero probability of ever happening. As summed up by the mathematician, Emil Borel (1962) "phenomena with very small probabilities do not occur."

Dr. Harold Klein, the chairman of a National Academy of Sciences committee which reviewed all the evidence, concluded that the simplest bacterium is so complicated it is impossible to imagine how it could have been created (Horgan 1991, p. 120). According to Dose (1988, p. 355), "The difficulties that must be overcome are at present beyond our imagination." Kuppers (1990, p. 60) sums it up this way: "The expectation probability for the nucleotide sequence of a bacterium is thus so slight that not even the entire space of the universe would be enough to make the random synthesis of a bacterial genome probable."

Given the complexity of DNA, and even a single protein, the likelihood that life could have arisen gradually and merely by chance, at least on Earth, is zero. The likelihood that a proto-organism may have been randomly created on Earth is zero. Adding to the completely improbability is the fact that many of the essential ingredients for DNA, RNA, nucleotide or protein construction were not available on this planet,

And, even if by some miracle these biological molecules had been constructed they would have been destroyed by the Hell that was Hadean Earth. Nevertheless, during this same 800 million year pummeling by extraterrestrial debris, not just Prokaryotes, but Eukaryotes were roaming the planet. Given the incredible complexity of a single protein, the greater complexity of a single nucleotide, and increasing complexity of a single DNA molecule then a single cell with its thousands of components, coupled with the complex genetic interactions and the variety of invading Prokaryotes which formed compartments, a nucleus, and mitochondria; and this took place randomly and according to Darwinian principles of evolution, in less than 800 million years and at a time when Earth was under a barrage of meteors, asteroids and comets, and when essential ingredients were not available; the very idea that life could have been fashioned on Earth via abiogenesis can only be described as based on incredible ignorance and magical thinking.

Beginning with a single gene, it takes at least 10 billion years to generate a minimal gene set sufficient to maintain the life of the simplest bacteria; and these estimates do not include the time necessary to produce the first proteins, then the first nucleotides, then the first DNA molecule, and then the first gene.

Life was present from the very beginning, as evident from biological activity discovered in the first rocks to re-solidify. Therefore, life must have arrived on Earth buried within the debris which was continually striking this planet. And, if Earth which became a member of this solar system as a much larger planet, had in fact been ejected from another solar system when its star entered the Red Giant phase, then it is also probable that microbial life living deep beneath the surface survived these conditions; some of the descendants of which migrated to the surface (Joseph 2009a). As detailed in earlier chapters, extremophiles and hyperthermophiles could easily survive and flourish on the Hell which was Earth.

Thus, despite the incessant pummeling of the planet and its hellish conditions, microbial life flourished, multiplied, and then these "genetic seeds of life" began to germinate, and then evolve according to basic genetic principles governing embryogenesis and metamorphosis. The implication, therefore, are that life on Earth came from other planets containing the genetic instructions for the generation of all manner of species of which life on Earth is merely a sample. Since these *genetic seeds* must have fallen upon innumerable worlds, then similar patterns of evolutionary embryogenesis and metamorphosis, and planetary terraforming accompanied by bursts of speciation, must have taken place on every Earth-like planet in the cosmos.

We Are Not Alone

Data from genetics, microbiology, astrobiology, astrophysics and cosmology, leads invariably to the conclusion that life has evolved on innumerable planets, including on worlds much older than our own. Some of these life forms must have been similar to those of Earth, particularly if they evolved on Earth-like planets.

What might be the nature of life on planets with a chemistry and environment completely unlike our own? If complex life exists on worlds billions of years older than Earth but on habitable planets similar to our own, how and in what way might it have evolved? Predictions can be made based on the evolution of life on this planet.

If humans do not become extinct and they continue to evolve, and if science marches on, then what might humans be like a million years from now? What about ten million, a hundred million, or a billion years from now? The old cortical brain of fish, frogs and reptiles does not contain neocortex, where humans the old cortical brain is enshrouded and covered with a thick neocortical mantle which makes language and rational thought possible. Might those who have evolved in the ancient corners of the cosmos have 10 layers of neocortex instead of six? Might the frontal lobes of the brain have greatly expanded conferring intellectual and perceptual abilities which completely dwarf our own? Cold they have genetically engineered their own evolution? There are stars which shine in the darkness of night which were born billions of years before the Earth was formed.

The Evidence for Extraterrestrial Life

In 1970 lunar soil samples were returned to Earth by the Luna 16 spacecraft in a hermetically sealed container and photographed (Rode et al., 1979). The photographs were later examined by Drs. Stanislav Zhmur, and Lyudmila M. Gerasimenko, who identified what they believed to be microfossils of coccoidal bacteria which resembled Siderococcus or Sulfolobus (Klyce, 2000; Zhmur & Gerasimenko, 1999). A third fossilized impression from the lunar surface resembles a spiral filamentous micro-Ediacaran (Joseph & Schild 2010a), a species which became extinct over 500,000 years ago.

In 1971, a TV camera from the lunar Surveyor Space Craft was retrieved by Apollo 12 astronauts, after sitting 3 years on the moon, and a single bacterium (Streptococcus mitis) was found within (Mitchell & Ellis, 1971). In addition, the lunar camera was discovered to be covered with a film of "organic material of unknown origin" (Flory & Simoneit, 1972; Simoneit & Burlingame, 1971). The possibility of contamination prior to sending the camera to the moon, or after it was returned, was ruled out by the scientists who made this discovery. It is impossible that the microbe was the result of some other form of contamination, such as a sneeze or cough. Since a droplet of saliva contains an average of 750 million organisms, if contamination of the lunar TV camera was due to a scientist's inadvertent cough or sneeze, a multitude of related bacteria, and a "representation of the entire microbial population would be expected," rather than a single species and a single organism (Mitchell & Ellis, 1971). Moreover, this Streptococcus mitis was dormant, but came back to life.

There is evidence of extant life on Mars as detected by the 1976 Viking Mission Labeled Release experiment, which exploited the sensitivity of 14C respirometry. Positive responses were obtained on the Viking 1 and 2 landing sites on Mars. The results indicated the possibility of living microorganisms on the red planet (Levin 2010; Levin & Straat 1976).

Microfossils and evidence of biological activity have been discovered in three meteors from Mars (Mckay et al. 1996; Thomas-Keprta et al., 2009).

In 2003, methane has been detected in the atmosphere of Mars, by the European Space Agency's Mars Express satellite.

Two separate teams of scientists have determined, based on a genomic analysis, that DNA-based life has a genetic ancestry leading backwards in time over 10 billion years (Joseph & Wickramasinghe 2011; Sharov & Gordon 2013), which is twice the age of Earth.

In 2011, and as based on computerized analysis of anomalous entities and structures filmed 200 miles above the planet by 10 NASA space shuttle missions, Joseph (2011) determined that these structures were plasma-like and engaged in life-like behaviors. Joseph reasoned that since extremophiles have been discovered in every conceivable environment on Earth, and as microbes can survive the conditions of space and have been recovered form the stratosphere, then why couldn't life live in the thermosphere which also part of Earth's atmosphere?

Beginning in the 1960s and continuing to the present, microfossils resembling various species of bacteria, including cyanobacteria have been repeatedly discovered in meteors older than this soler system (Claus and Nagy 1961; Hoover 1998, 2006, 2011; Pflug 1984; Nagy et al. 1961,1963a,b; Zhmur & Gerasimenko 1999; Zhmur et al. 1997). Moreover, fossils of Cyanobacteria colonies were discovered in the Murchison meteor (Hoover, 2011) which is older than this solar system.

Coupled with the genetic evidence reviewed in this text, and given that life could not have begun on Earth, it is thus reasonable to conclude that life on Earth came from other planets, and that the cosmos is flush with life.

Cyanobacteria and the Genetic Seeds of Life

The discoveries of microfossils Cyanobacteria and colonies of these microbes in the Murchison meteor (Hoover, 2011) is of particular importance as it is these Prokaryotes which biologically engineered Earth by creating an oxygen atmosphere which generate a UV protecting ozone layer, and calcium which is utilized for generating shells, bones and brains (Joseph 2000a, 2009d, 2010a).

Cyanobacteria are a hardy species, and can live in extreme environments. Therefore, when Cyanobacteria are deposited on Earth-like planets, it can be assumed they also biologically engineered these alien worlds, providing them with an oxygen atmosphere and flooding the environment with calcium and inserting photosynthesizing and calcium producing genes in other species.

Coupled with horizontal gene transfers and the crucial role of Prokaryotes and viruses in providing the Eukaryotic genome with core genetic material including regulatory genes, it can be assumed that when this assemblage of microbes fell upon other Earth-like worlds that identical genetic interactions took place. Thus, on habitable planets it can be deduced that these "genetic seeds of life" fell upon innumerable planets thereby making it possible for the evolution of bones and brains and for life to evolve into intelligent species, similar to or completely different from, and possibly more intelligent than woman and man.

And what if these bacterial "seeds of life" fell upon planets unlike our own? If they could take root and flourish, they might evolve into creatures completely unlike those of Earth. This might account for the truly "alien" microbes discovered by Hoover (2011).

It is also well established the microbes and viruses can withstand a journey through space, including the ejection from and crashing on the surface of planet, and the radioactive hazards they may encounter while sojourning through space, so long as they have minimal shielding. Through powerful solar winds microbes living in the upper atmosphere of Earth or alien planets may be cast into space, and then there are bolide impacts which can eject chunks of various planets into space; and if those planets harbor life, then microbes living in those ejected rocks, bodies of water, or mounds of soil would likely survive. And once the survivors fall upon the surface of another planet, they may go forth and multiply.

The implications are profound. It can be assumed that life is everywhere and has a cosmic ancestry which extends backwards in time, interminably into the long ago, and that intelligent life has evolved on countless Earth-like planets (Joseph 2000a, 2010c). And we can predict that life must have continued to evolve on innumerable worlds which are much older than Earth.

But what forms might they take? Might they be human or could they be intelligent plants or insects that ask: "are we alone?"

Evolution of the Future: Even the Gods Have Gods...

Anatomically "modern" humans (such as the Cro-Magnon people) have resided on Earth for less than 50,000 years. In the last 300 years scientific and technological advances have exploded exponentially. It is reasonable to ask, if humans do not self-destruct and are not forced into extinction, and if scientific and technological progress continues, what might humans be like in another 50,000 years? There is no reason to believe evolution will stop with modern humans.

Alien Viruses, Genetics, Cambrian Explosion: Humans

It can be predicted that humans will eventually harness the ability to genetically modify and design babies with superior brains and bodies; perhaps within the next 500 years. If these designer babies design their own babies, and if science marches on, if evolution is an ongoing process of metamorphosis, what might humans be like in a million years? What about in 10 million? Or a hundred million? Or a billion? From the perspective of modern humans, they may appear as gods even if they are still human. Even the gods may have gods.

There are stars which twinkle in the darkness of night which are 13 billion years in age, many of which are likely orbited by habitable planets where life began to evolve 9 billion years before Earth became Earth. What might be the intellectual capabilities of advanced life forms which have been evolving for over 10 billion years? They too would appear as gods with intellectual and scientific powers completely beyond our comprehension.

And if the creationists are wrong and there was no big bang, if the universe is eternal and infinite, then life, too, has had infinite time to evolve on infinite worlds. All manner of life, including those which were never human, may have evolved beyond modern human understanding infinitely long ago.

Even the "gods" may have "gods" who have "gods" who have "gods"... an infinity of genetic potentiality which may be represented in the genomes of extraterrestrial Prokaryotes, innumerable viral species, and "modern" humans.

Life on Earth, came from other planets. Evolution is metamorphosis: Evolutionary metamorphosis. Evolutionary embryology.

Life on Earth is continuing to develop, to undergo metamorphosis, and evolve. There is every reason to suspect that evolution and metamorphosis will not cease with humans, but will continue, as dictated and determined by the genetic seeds of life which swarm throughout the cosmos, replicating with each passing epoch, life forms which long ago evolved on other planets.

Yes, our children's children's children's children... may become as "gods" even though they may still be human. And this is because:

Life on Earth Came From Other Planets
Our Ancient Ancestors, Journeyed Here, From the Stars.

10. REFERENCES

Abbas, S., Abbas, A. & Mohanty, S. (2000). Anoxia during the Late Permian binary mass extinction and dark matter. Current Science, 78, 30-33.

Abyzov, S. et al., (1998). Microbiologiya, 67, 547.

Ackermann HW, et al., (1987). Viruses of prokaryotes: General properties of bacteriophages. Boca Raton, Florida: CRC Press, Inc.

Ackermann HW (2007). 5500 Phages examined in the electron microscope. Arch Virol 2007, 152:227-243.

Ackermann HW, DuBow MS. (1987). Viruses of prokaryotes: General properties of bacteriophages. CRC Press, Inc

Adamowicz, S. J., Purvis, A. & Wills, M. A. (2008) Increasing morphological complexity in multiple parallel lineages of the Crustacea. Proc. Natl Acad. Sci. USA 105, 4786–4791

Agrawal A, Eastman QM, Schatz DG (1998) Transposition mediated by RAG1 and RAG2 and its implications for the evolution of the immune system. Nature 394: 744Y751.

Akao, M., et. al., (2001). Mitochondrial ATP-Sensitive Potassium Channels Inhibit Apoptosis Induced by Oxidative Stress in Cardiac Cells Circulation Research. 88, 1267-1275.

Alibert, Y., Baraffe, I., Benz, W., et. al. (2006). Formation and structure of the three Neptune-mass planets system around HD 69830. Astronomy & Astrophysics, 455, L25.

Alvarez. W. (2003) Comparing the Evidence Relevant to Impact and Flood Basalt at Times of Major Mass Extinctions. Astrobiology. January 2003, 3, 153-161.

Alvarez, W. (2008). "T. rex" and the Crater of Doom, Princeton Science Library.

Alvarez, L. W., Alvarez, W., Asaro, F., and Michel, H. V. (1980). Extraterrestrial cause for the Cretaceous-Tertiary extinction. Science, 208: 1095-1108.

Alvarez-Buylla E.R., et al., (2000). An ancestral MADS-box gene duplication occurred before the divergence of plants and animals. Proc Natl Acad Sci U S A. 9;97(10):5328-3

Alvarez-Ponce, D., Lopez, P., Bapteste, E., McInerney, J. O. (2013) Gene similarity networks provide tools for understanding eukaryote origins and evolution, PNAS 2010 107 (27) 12168-12173

Amaral et al. (2008). The Eukaryotic Genome as an RNA Machine, Science, 319. 1787 - 1789.

Anders, E. (1989). Nature 342, 255.

Anderson. S, et al. (1981). "Sequence and organization of the human mitochondrial genome". Nature. 410. 141.

Andersson J.O. (2005). Lateral gene transfer in eukaryotes., Cell Mol Life Sci. 62(11):1182-97.

Andersson A-C, Venables PJW, Tönjes RR, et al. (2002). Developmental expression of HERV-R (ERV-3) and HERV-K in human tissue. Virology, 297:220-225.

Anisimov, V. (2010). Principles of Genetic Evolution and the ExtraTerrestrial Origins of life. Journal of Cosmology, 5, 843-850.

Appasania, K (2012). Epigenomics, From Chromatin Biology to Therapuetics, Cambridge U. Press.

Aravind L, Tatusov RL, Wolf YI, Walker DR, Koonin EV. (1998). Evidence for massive gene exchange between archaeal and bacterial hyperthermophiles. Trends Genet. 14:442–444.

Aravind L, et al. (1999). The domains of death: evolution of the apoptosis machinery. Trends Biochem. Sci. 24:47–53.

Aravind, L., et al (2000). Lineage-specific loss and divergence of functionally linked genes in eukaryotes. Proc. Natl. Acad. Sci. 97: 11319-11324.

Arber, W., & Linn, S. (1969). DNA modification and restriction. Annual Review of Biochemistry 38, 467–500.

Arbeitman, MN and DS Hogness. 2000. Molecular chaperones activate the Drosophila ecdysone receptor, an RXR heterodimer. Cell 101:67–77.

Arendt, D. and Nübler-Jung, K. (1999). Comparison of early nerve cord development in insects and vertebrates. Development 126, 2309-2325.

Arens, N. C. and West, I. D. (2008). Press-pulse: a general theory of mass extinction? Paleobiology, 34 (4): 456-471.

Arnett, D. (1996) Supernovae and Nucleosynthesis. Princeton University Press.

Arrhenius, S. (2009). The Spreading of Life Throughout the Universe. Journal of Cosmology, 2009, 1, 91-99.

Ash R. D., Knott S. F., and Turner G. (1996) A 4-Gyr shock age for a martian meteorite and implications for the cratering history of Mars. Nature, 380, 57-59.

Ayala, F. J., Rzhetsky, A. and Ayala, F. J. (1998). Origin of the metazoan phyla: Molecular clocks confirm paleontological estimates. Proc. Natl. Acad. Sci. USA 95, 606-611.

Arouri, K. R. , et al., (2000). WalteraBiological affinities of Neoproterozoic acritarchs from Australia: microscopic and chemical characterisation, Organic Geochemistry, 31,75-89.

Arrhenius, S. 1908. Worlds in the Making. Harper & Brothers, New York.

Babenko, V.N., Rogozin, I.B., Mekhedov, S.L., Koonin, E.V. (2004) Prevalence of intron gain over intron loss in the evolution of paralogous gene families. Nucleic Acids Res. 32:3724–3733.

Bailey, J., et al., (1998). Circular Polarization in Star- Formation Regions: Implications for Biomolecular Homochirality. Science 31 July 1998: Vol. 281. no. 5377, pp. 672 - 674.

Bakatselou C, Beste D, Kadri AO, Somanath S, Clark CG (2003). Analysis of genes of mitochondrial origin in the genus Entamoeba. J. Eukaryotic Microbiology 50, 210–214.

Baker, ME. (1997). Steroid receptor phylogeny and vertebrate origins. Mol Cell Endocrinol 135:101–7.

Baker, ME. (2003). Evolution of adrenal and sex steroid action in vertebrates: a ligand-based mechanism for complexity. BioEssays 25:396–400.

Baker, ME. (2005). Xenobiotics and the evolution of multicellular animals: emergence and diversification of ligand-activated transcription factors. Integr Comp Biol 45:172–8

Baker, ME. (2006). Evolution of metamorphosis: role of environment on expression of mutant nuclear receptors and other signal-transduction proteins. Integrative and Comparative Biology 2006 46(6):808-814.

Bakos, G.A , Kovacs, G., Torres, G., et al. (2007), ApJ, astro-ph/7050126.

Bao, H. et al. (2008). Triple oxygen isotope evidence for elevated CO2 levels after a Neoproterozoic glaciation. Nature 453, 504-506.

Bapst, D. W., Bullock, P. C., Melchin, M. J., Sheets, H. D. & Mitchell, C. E. (2012) Graptoloid diversity and disparity became decoupled during the Ordovician
mass extinction. Proc. Natl Acad. Sci. USA 109, 3428–3433 .

Baraffe, G., Chabrier, T. Barman, G. (2008). Structure and evolution of super-Earth to super-Jupiter exoplanets: I. heavy element enrichment in the interior. Astronomy & Astrophysics, 482, 315 - 332.

Barbulescu M, Turner G, Seaman MI, Deinard AS, Kidd KK, Lenz J. (1999). Many human endogenous retrovirus K (HERV-K) proviruses are unique to humans. Curr Biol. 9(16):861-8.

Barleya, M. E., et al., (2005). Late Archean to Early Paleoproterozoic global tectonics, environmental change and the rise of atmospheric oxygen--Earth and Planetary Science Letters 238, 156-171.

Barlow, N. (1959). The Autobiography of Charles Darwin (1959).W.W. Norton & Co.

Bartoloni A., et al., (2009). Antibiotic resistance in a very remote Amazonas community. Int. J. Antimicrob. Agents. 33, 125–129.

Battistuzzi, F. U. and Hedges, S. B. (2009). A Major Clade of Prokaryotes with Ancient

Adaptations to Life on Land," Mol. Biol. Evol. 26 (2), 335–343.

Baymann, F. (2003). The redox protein construction kit: pre-last universal common ancestor evolution of energy-conserving enzymes. Philos Trans R Soc Lond B Biol Sci. 358(1429): 267-274.

Beare MH, Parmelee RW, Hendrix PF, Cheng W (1992) Microbial and faunal interactions and effects on litter nitrogen and decomposition in agroecosystems. Ecological Monographs 62: 569-591.

Bejerano, G., (2004). Ultraconserved Elements in the Human Genome Science, 304. 1321 - 1325.

Belbruno, E., Gott III, J. R. (2005). Where Did the Moon Come From? The Astronomical Journal 129 1724-1745.

Belfort, M. (1991). Self-splicing introns in prokaryotes, Cell, 64, 9-11.

Belfort, M. (1993). Introns. Science, 262, 1009-1010.

Belloche, A., Garrod, R.T., Muller, H.S.P.. Menten, K.M., Comito, C., and Schilke, P. (2009). Increased Complexity in Interstellar Chemistry : Detection and Chemical Modelling of Ethyl Formate and n-propyl Cyanide in Sagittarius B(2) N. Astronomy and Astrophysics, 499, 215.

Belshaw R, Katzourakis A, Paces J, Burt A, Tristem M (2005) High copy number in human endogenous retrovirus families is associated with copying mechanisms in addition to reinfection. Mol Biol Evol 22: 814Y817.

Benett, C.L. et al., (2003). Astrophy. J. Suppl. 148, 1-27.

Bensasson, D., et al. (2001). Mitochondrial pseudogenes: evolution's misplaced witnesses, Trends in Ecology & Evolution, e 16, 314-321.

Benton, M. J., & Twitchett, R/. J., (2003). How to kill (almost) all life: the end-Permian extinction event Trends in Ecology & Evolution, 18, 358-365.

Berger, B., et al., (2013). Chimerism in DNA of buccal swabs from recipients after allogeneic hematopoietic stem cell transplantations: implications for forensic DNA testing, International Journal of Legal Medicine, 127, 49-54.

Berkner, K. L. (1988). Development of adneovirus vectors for the expression of heterologous genes. Biochemical Techniques, 6, 616-629.

Berks, B. C., Page, M. D., Richardson, D. J., Reilly, A., Cavill, A., Outen, F. & Ferguson, S. J. (1995). Sequence analysis of subunits of the membrane-bound nitrate reductase from a denitrifying bacterium: the integral membrane subunit provides a prototype for the dihaem electron-carrying arm of a redox loop.Mol Microbiol 15, 319-331.

Bernstein, G. 2004, Hubble Space Telescope Newsletter, v 21, #1, Winter 2004, p.18; see also astro-ph/0308.467.

Biesecker, L. G. & Spinner, N. B. (2013) A genomic view of mosaicism and human disease, Nature Genetics, 14, 307-320

Blaise S, de Parseval N, Benit L, Heidmann T (2003) Genomewide screening for fusogenic human endogenous retrovirus envelopes identifies syncytin 2, a gene conserved on primate evolution. Proc Natl Acad Sci USA 100: 13013Y13018.

Blond J-L, Lavillette D, Cheynet V, et al. (2000). An envelope glycoprotein of the human endogenous retrovirus HERV-W is expressed in the human placenta and fuses cells expressing the type D mammalian retrovirus receptor. J Virol; 74:3321 -3329.

Boeke, J. D., Stoye, J. P. (1997). Retrotransposons, endogenous retroviruses, and the evolution of retroelements, p. 343-435. In J. M. Coffin, S. H. Hughes, and H. E. Varmus (ed.), Retroviruses. Cold Spring Harbor Laboratory Press, New York, N.Y.

Böhne, A., et al., (2008). Transposable elements as drivers of genomic and biological diversity in vertebrates Chromosome Research, 16, 21-33.

Boone, D R., Liu, Y., Zhao, Z. J., Balkwill, D. L., Drake, G. R., Stevens, T. O., Aldrich, H. C. (1995). Bacillus infernus sp. nov., an Fe(III)- and Mn(IV)-reducing anaerobe from the deep terrestrial subsurface. International journal of systematic bacteriology. 45(3):441-8.

Borel, E. (1962). Probability and Life, Dover.

Bornemann, A., et al. (2008). Isotopic Evidence for Glaciation During the Cretaceous Supergreenhouse Science, 319, 189 - 192.

Bose, R. (2013). Morphological Evolution in an Atrypid Brachiopod Lineage from the Middle Devonian Traverse Group of Michigan, USA: A Geometric Morphometric Approach, Biodiversity and Evolutionary Ecology of Extinct Organisms, 41-62

Bowring, S. A., et al. (2003) Geochronological constraints on terminal Neoproterozoic events and the rise of Metazoans. NASA Astrobiol Inst (NAI Gen Mtg Abstr) 2003:113–114.

Bradshaw, C.J.A., Brook, B.W. (2009). The Cronus Hypothesis: Extinction as a Necessary and Dynamic Balance to Evolutionary Diversification. Journal of Journal of Cosmology, 2009, 2, 221-229.

Breitbart, R. E., Nguyen, H. T., Medford, R. M., Destree, A. T., Madhavi, V. & Nadal-Ginard, B. (1985). Cell 41, 67-82.

Brenchley, P. J., Marshall, J. D., & Underwood, C. J. (2001). Do all mass extinctions represent an ecological crisis? Evidence from the Late Ordovician. Geological Journal, 36, 329 - 340.

Brochier C, Philippe H, Moreira D. (2000). The evolutionary history of ribosomal protein RpS14: horizontal gene transfer at the heart of the ribosome. Trends Genet. (2000) 16:529–533.

Brocks, J.J., Love G.D, Summons R.E, Knoll A.H, Logan G.A, Bowden S.A (2005). Biomarker evidence for green and purple sulphur bacteria in a stratified Palaeoproterozoic sea. Nature. 437, 866–870.

Brown, N. F., et al., (2006). Crossing the line: selection and evolution of virulence traits. PLoS Pathog. 2006 May ;2 (5):e42 16733541.

Brownlee, D. 2008, Physics Today, June 2008, p. 30

Brodie, E. L., DeSantis, t. Z., Parker, J. P. M., Zubietta, I. X., Piceno, Y. M., Andersen, G. L. 2007. Urban aerosols harbor diverse and dynamic bacterial populations PNAS. 104, 299-304.

Brown, J. R., & Doolittle, W. F. (1997). Archaea and the prokaryote-to-eukaryote transition Microbiol Mol Biol Rev. 61, 456–502.

Brussow H, Canchaya C, Hardt WD. (2004). Phages and the evolution of bacterial pathogens: From genomic rearrangements to lysogenic conversion. Microbiology and Molecular Biology Reviews. 68:56-566.

Budd, G.E., & Jensen, S. (2000). A critical reappraisal of the fossil record of the bilaterian phyla. Biol Rev. 75, 253–95.

Buick, R. (1992). The antiquity of oxygenic photosynthesis: evidence from stromatolites in sulphate-deficient Archaean lakes Science, Vol 255, Issue 5040, 74-77.

Buick, R., (2008). When did oxygenic photosynthesis evolve?--Phil. Trans. R. Soc. B 27 363 no. 1504 2731-2743.

Bult, C. J., White, O., Olsen, G. J., et al., (1996). Complete genome sequence of the methanogenic archaeon, Methanococcus jannaschii. Science 273, 1058–1073.

Burchell, J. R. Mann, J., Bunch, A. W. (2004). Survival of bacteria and spores under extreme shock pressures, Monthly Notices of the Royal Astronomical Society, 352, 1273-1278.

Burchella, M. J., Manna, J., Bunch, A. W., Brandãob, P. F. B. (2001). Survivability of bacteria in hypervelocity impact, Icarus. 154, 545-547.

Burbidge, E. M. Burbidge, G. R.. Fowler, W. A. Hoyle, F. (1957). Synthesis of the Elements in Stars, Rev. Mod. Phys. 29.

Burrows, A., Hubeny, I., Budaj, J., Hubbard, W.B. (2007), ApJ, 661, 502.

Butterfield, N. J. (2000). Bangiomorpha pubescens n. gen., n. sp.: implications for the evolution of sex, multicellularity, and the Mesoproterozoic/Neoproterzoic radiation of eukaryotes. Paleobiology, 26, 386-404.

Butterfield, N.J (2005a). Probable Proterozoic fungi. Paleobiology. 31, 2005a 165–182.

Butterfield, N.J. (2005b). Reconstructing a complex early Neoproterozoic eukaryote, Wynniatt formation, arctic Canada. Lethaia. 38, 155–169.

Butterfield, N. J. (2000). Bangiomorpha pubescens n. gen., n. sp.: implications for the evolution of sex, multicellularity, and the Mesoproterozoic/Neoproterzoic radiation of eukaryotes. Paleobiology 26, 386-404.

Buzdin A, Ustyugova S, Khodosevich K, Mamedov I, Lebedev Y, Hunsmann G, Sverdlov E. (2003). Human-specific subfamilies of HERV-K (HML-2) long terminal repeats: three master genes were active simultaneously during branching of hominoid lineages. Genomics, 81(2):149-56.

Caballero, J. et al., (2007), A&A, 470, 903.

Callaerts, P., Halder,G., Gehring, W. J., (1997). Pax-6 in development and evolution. Annual Review of Neuroscience, 20: 483-532.

Cairns J, Stent GS, Watson JD, eds. (1966). Phage and the Origins of Molecular Biology (1966) Cold Spring Harbor, NY: CSHL Press.

Canfield, D.E. (2005). The early history of atmospheric oxygen. Annu. Rev. Earth Planet. Sci. 33, 1–36.

Carmel, L., et al., (2007). Three distinct modes of intron dynamics in the evolution of eukaryotesGenome Res. 17: 1034-1044.

Carr, J. S., and Najita, J. R. (2008). Organic Molecules and Water in the Planet Formation Region of Young Circumstellar Disks. Science. 319. 1504 - 1506.

Carlson, B. M., (2013) Human Embryology and Developmental Biology, Saunders.

Casadevall, A. (2005). Fungal virulence, vertebrate endothermy, and dinosaur extinction: is there a connection? Fungal Genetics and Biology, 42, 98-106.

Casagrande, V. A. & Joseph, R. (1978). Effects of monocular deprivation on geniculostriate connections in prosimian primates. Anatomical Record, 190, 359.

Casagrande, V. A. & Joseph, R. (1980). Morphological effects of monocular deprivation and recovery on the dorsal lateral geniculate nucleus in prosimian primates. Journal of Comparative Neurology, 194, 413-426.

Castle, J. and Rodgers, J. (2009) Hypothesis for the role of toxin-producing algae in Phanerozoic mass extinctions based on evidence from the geologic record and modern environments. Environmental Geosciences 16 (1): 1-23.

Castresana, J. & Moreira, D. (1999). Respiratory chains in the last common ancestor of living organisms. J Mol Evol 49, 453-460.

Cavalier-Smith, T. (1991). Intron phylogeny: a new hypothesis. Trends Genet. 7, 145–148.

Cavalier-Smith, T., (2006). Cell evolution and Earth history: stasis and revolution--Phil. Trans. R. Soc. B 29, 361, 969-1006.

Cerrigone, L. et al., (2009). Spitzer Detection of Polycyclic Aromatic Hydrocarbons and Silicate Features in Post-AGB Stars and Young Planetary Nebulae. The Astrophysical Journal, 703, 585-600.

Chakrabarti A C; Deamer D W. Permeability of lipid bilayers to amino acids and phosphate. Biochimica et biophysica acta 1992;1111(2):171-7.

Charlebois RL, Doolittle WF. (2004). Computing prokaryotic gene ubiquity: rescuing the core from extinction. Genome Res. 14:2469–2477.

Chen, K., et al. (2013). Chimerism in monochorionic dizygotic twins: Case study and review, American Journal of Medical Genetics, 161, 1817-1824.

Chimpanzee Sequencing and Analysis Consortium (2005) Initial sequence of the chimpanzee genome and comparison with the human genome. Nature 437: 69Y87.

Chipuk, J.E., Bouchier-Hayes, L., Green, D.R. (2006). Mitochondrial outer membrane permeabilization during apoptosis: the innocent bystander scenario. Cell Death and Differentiation. 13: 1396–1402.

Chivian, D., et al., 2008. Environmental Genomics Reveals a Single-Species Ecosystem Deep Within Earth Science 322, 275-278.

Chou, M-I. et al., (2009). A Two Micron All-Sky Survey View of the Sagittarius Dwarf Galaxy. http://arxiv.org/pdf/0911.4364.

Ciftcioglu, N. et al.., (2006). Nanobacteria: Fact or Fiction? Characteristics, Detection, and Medical Importance of Novel Self-Replicating, Calcifying Nanoparticles Journal of Investigative Medicine, 54, 385-394.

Claus, G., Nagy, B. (1961) A Microbiological Examination of Some Carbonaceous Chondrites. Nature 192, 594 - 596.

Clark, R. N. (2009). Detection of Adsorbed Water and Hydroxyl on the Moon. Science, 326, 562 - 564.

Claus, G., Nagy, B. (1961) A Microbiological Examination of Some Carbonaceous Chondrites. Nature 192, 594 - 596.

Clayton, D. (1984) Principles of Stellar Evolution and Nucleosynthesis. University Of Chicago Press.

Clayton, D. D. (2003). Handbook of Isotopes in the Cosmos", Cambridge University Press.

Clayton, R. N. 2002, Solar system: Self-shielding in the solar nebula, Nature 415, 860-861.

Claverie JM. (2005). Giant viruses in the oceans: the 4th Algal Virus Workshop. Virol J. 20;2:52.

Cloud, P.E. (1948). Some problems and patterns of evolution exemplified by fossil invertebrates. Evolution, 2, 322–350.

Collins, L., Penny, D. (2005) Complex spliceosomal organization ancestral to extant eukaryotes. Mol. Biol. Evol. 22:1053–1066.

Comeron, J. M. & Kreitman, M. (2000). The correlation between intron length and recombination in Drosophila. Dynamic equilibrium between mutational and selective forces. Genetics 156, 1175–1190.

Condon,D., et al. (2005). U-Pb Ages from the Neoproterozoic Doushantuo Formation, China Science, 308, 95 - 98.

Conley, A. B., Piriyapongsa, J., Jordan, I. K. (2008). Retroviral promoters in the human genome, Bioinformatics, 24, 1563-1567.

Conway Morris, S. (2000). The Cambrian "explosion": slow-fuse or megatonnage? Proc Natl Acad Sci. 97, 4426–4429.

Cossins, A. (1998). Cryptic clues revealed. Nature, 396, 309-310.

Costas, J. (2001). Evolutionary dynamics of the human endogenous retrovirus family HERV-K inferred from full-length proviral genomes. J. Mol. Evol. 53:237-243.

Courtillot, V. (1999). Evolutionary Catastrophes: The Science of Mass Extinction. Cambridge University Press.

Couvault, C. (2006). Aviation Week and Space Technology, March 20, 2006, p.31.

Cowan, M.K., Talaro, K. P., (2008) Microbiology: A Systems. Approach. McGraw-Hill Science.

Cox, T. J., and Loeb, A. (2008). The Collision Between The Milky Way And Andromeda, Monthly Notices of the Royal Astronomical Society, 386, 461-474.

Crick, F. (1981). Life Itself. Its Origin and Nature. Simon & Schuster, New York.

Crombach A, Hogeweg P. (2007). Chromosome Rearrangements and the Evolution of Genome Structuring and Adaptability. Mol Biol Evol. ;24:1130–1139.

Crutzen, P.J., Stoermer, E.F. (2000). The "Anthropocene". Global Change Newsletter, 41, 12-13.

Csuros M, Rogozin IB, Koonin EV. (2008). Extremely intron-rich genes in the alveolate ancestors inferred with a flexible maximum-likelihood approach. Mol. Biol. Evol. 25:903–911.

D'Acosta V.M., McGrann K.M., . Hughes D.W., Wright G.D.(2006). Sampling the antibiotic resistome. Science. 311, 374–377.

Dahout-Gonzalez, C., Nury, H., Trézéguet, V., Lauquin, G., Pebay-Peyroula, E., Brandolin, G. (2006). Molecular, functional, and pathological aspects of the mitochondrial ADP/ATP carrier. Physiology, 21, 242–249.

Dai L, Zimmerly S. (2003). ORF-less and reverse-transcriptase-encoding group II introns in archaebacteria, with a pattern of homing into related group II intron ORFs. RNA.9 :14-9.

Dai L, Toor N, Olson R, Keeping A, Zimmerly S. (2003). Database for mobile group II introns. Nucleic Acids Res. 31:424-6.

Dai J, Xie W, Brady TL, Gao J, Voytas DF. (2007). Phosphorylation regulates integration of the yeast Ty5 retrotransposon into heterochromatin. Mol Cell. 27, 289-99.

Dantas G., et al. (2008). Bacteria Subsisting on Antibiotics Science, 320, 100 - 103.

Darwin, C. (1838) Notebook C: Transmutation (1838)

Darwin, C. (1857). Letter to Asa Gray, 18 June, 1857.

Darwin, C. (1859), letter to Sir Charles Lyell, November 23, 1859.

Darwin, C. (1859). The origin of species by means of natural selection. London, Murray.

Darwin, C. (1866) Origin of Species: fourth British edition (1866)

Darwin, C. (1871). The origin of species and the descent of man. New York, Random House.

Darwin, C. (1887). Letters. In Darwin, F. (ed.), The Life and Letters of Charles Darwin, Vols. 1 & 2. Appleton, New York.

Darwin, C. (1892). Autobiography. The Autobiography of Charles Darwin 1809-1882

Dassa B, Amitai G, Caspi J, Schueler-Furman O, Pietrokovski S. (2007). Trans protein splicing of cyanobacterial split inteins in endogenous and exogenous combinations. Biochemistry 46:322–330.

Davidson, E. H. (2001). Genomic Regulatory Systems. Development and Evolution. San Diego: Academic Press.

Davies J. (1994). Inactivation of antibiotics and the dissemination of resistance genes. Science. 264, 375–382.

Davies J.E. (1997) Origins, acquisition and dissemination of antibiotic resistance determinants. Ciba Found. Symp. 207, 15–27.

Day M, (1998) in Horizontal Gene Transfer, eds Syvanen M, Kado C I (Chapman & Hall, London), pp 144–167.

Davidson, E. H. (2001). Genomic Regulatory Systems. Development and Evolution. San Diego: Academic Press.

Dayhoff MO, Barker WC, McLaughlin PJ. (1974). Inferences from protein and nucleic acid sequences: early molecular evolution, divergence of kingdoms and rates of change. Orig. Life. 5:311–330.

Dayhoff MO, Barker WC, Hunt LT. (1983). Establishing homologies in protein sequences. Methods Enzymol. 91:524–545.

(De Coppi, P., et al., (2007) Isolation of amniotic stem cell lines with potential for therapy. Nature Biotechnology 25, 100 - 106.

de Duve C. and Osborn M. J. (1999) Panel 1: Discussion. In Size Limits of Very Small Microorganisms: Proceedings of a Workshop. (National Academy Press, Washington, DC).

Deininger, P. L. & Batzer, M. A. (2002) Mammalian Retroelements. Genome Res. 12 , 1455-1465.

Delgado-Iribarren A.,Martinez-Suarez J., Baquero F., Perez-Diaz J.C., Martinez J.L. (1987). Aerobactin-producing multi-resistance plasmids. J. Antimicrob. Chemother. 19, 552–553.

Dehal, P., & Boore, J.L.. (2005). Two rounds of whole genome duplication in the ancestral vertebrate. PLoS Biol. 3, 314.

De Jong, G., & Scharloo, W. (1976). Environmental determination of selective significance of neutrality of amylase variants in Drosophilia. Genetics 84, 77-94.

de Koning, A. P., et al., (2000). Lateral Gene Transfer and Metabolic Adaptation in the Human Parasite Trichomonas vaginalis Molecular Biology and Evolution 17:1769-1773.

De Rosa M, Gambacorta A, Gliozzi A (1986). "Structure, biosynthesis, and physicochemical properties of archaebacterial lipids". Microbiol. Rev. 50 (1): 70–80.

De Souza, S. J. (2003). The emergence of a synthetic theory of intron evolution. Genetica 118, 117–121.

De Souza, S. J. et al. (1998). Towards a resolution of the introns early/late debate: only phase zero introns are correlated with the structure of ancient proteins. Proc. Natl Acad. Sci. USA 95, 5094–5099.

De Souza, S. J., Long, M., Schoenbach, L., Roy, S. W. & Gilbert., W. (1996). Introns correlate with module boundaries in ancient proteins. Proc. Natl Acad. Sci. USA 93, 14632–14636.

De Souza, S. J., Long, M., Gilbert, W. (1996). Introns and gene evolution. Genes to Cells, 1 495-505.

Devaraj, M.S., (2000). Mechanisms of Extinction-Viruses. Journal of Theoretics, Vol.2-3, 1-23.

Dhimolea, E. et al. (2013). High male chimerism in the female breast shows quantitative links with cancer, International Journal of Cancer, 133, 835-842.

Dieci, G., Preti, M., Montanini, B. (2009) Eukaryotic snoRNAs: A paradigm for gene expression flexibility, Genomics, 94, 83-88.

Diemand, J. Moore, B. and Stadel, J. (2005). Nature, 433, 389 .

Dietrich, M., et al., (2009) Black Hole Masses of Intermediate-Redshift Quasars: Near Infrared Spectroscopy. The Astrophysical Journal, 696, 1998-2013.

Doerfert, S. N. (2009). Methanolobus zinderi sp. nov., a methylotrophic methanogen isolated from a deep subsurface coal seam. Int J Syst Evol Microbiol 59, 1064-1069.

Dombrowski, H. (1963). Bacteria from Paleozoic salt deposits. Annals of the New York Academy of Sciences, 108, 453.

Doolittle, W.F. (1978) Genes in pieces: Were they ever together? Nature 272:581–582.

Doolittle, W.F. (1999). Phylogenetic classification and the universal tree. Science 284, 2124–2129.

Doolittle W F, Sapienza C (1980) Selfish genes, the phenotype paradigm and genome evolution Nature 284:601–603.

Doolittle RF, Feng DF, McClure MA, Johnson MS. (1990). Retrovirus phylogeny and evolution.Curr Top Microbiol Immunol. 157, 1-18.

Dose, K. (1988). The origin of life: More questions than answers. Interdisciplinary Science Review, 13, 348-356.

Drake, M. J. (2005). Origin of water in the terrestrial planets, Meteoritics & Planetary Science, 40, 515–656.

Druschke, P. et al. (2006). Stratigraphic and U-Pb SHRIMP Detrital Zircon Evidence for a Neoproterozoic Continental Arc, Central China: Rodinia Implications The Journal of Geology, 114, 627–636.

Du, R. Q., Jin, L., Zhang, F. (2011). Copy number variations in the human genome: their mutational mechanisms and roles in diseases, Yi Chuan, 33, 857-869.

Dufour, S., Rousseau. K., Kapoor, B. G. (2012) Metamorphosis in Fish, Science Publishers.

Duret, L. (2001). Why do genes have introns? Recombination might add a new piece to the puzzle. Trends Genet. 17, 172–175.

Dyall SD, Brown MT, Johnson PJ (2004) Ancient invasions: From endosymbionts to organelles. Science 304: 253–257.

Dyall, S. D, Johnson, P. J. (2000). Origins of hydrogenosomes and mitochondria: evolution and organelle biogenesis-- 1. Current Opinion in Microbiology, 3, 404-411.

Dykhuizen, D., & Hart, D. L. (1980) Selective neutrality of 6PDG alozymes in E. coli and the effects of genetic background. Genetics 96, 801-817. Eck RV, Dayhoff MO. (1966). Evolution of the structure of ferredoxin based on living relics of primitive amino acid sequences. Science 152:363–366.

Ehrenfreund. P. & Menten, K. M. (2002). From Molecular Cluds to the Origin of Life. In G. Horneck & C. Baumstark-Khan. Astrobiology, Springer.

Ehrenfreund. P., and Sephton, M. A. (2006). Carbon molecules in space: from astrochemistry to astrobiology. Faraday Discuss., 2006, 133, 277 - 288.

Elbaz. D., et al., (2009) Quasar induced galaxy formation: a new paradigm ? Astronomy and Astrophysics, 507, 1359-1374.

Elewa, A. M. T. (2008). Mass Extinction – A General View. In Elewa A. M. T. (ed.): Mass Extinction. Springer–Verlag Publishers, Heidelberg, Germany, 1-6.

Elewa, A. M. T. & Joseph, R. (2009). The History, Origins, and Causes of Mass Extinctions, Journal of Cosmology, 2,201-220.

Elvis, M., et al., (1994). Atlas of quasar energy distributions. The Astrophysical Journal Supplement Series, 95, 1-68.

Eldredge, N., Gould, S. J., (1972). Punctuated equilibria: an alternative to phyletic gradualism" In T.J.M. Schopf, ed., Models in Paleobiology. San Francisco: Freeman Cooper. pp. 82-115.

Elewa, A. M. T., and Joseph, R. (2009). The History, Origins, and Causes of Mass Extinctions. Journal of Cosmology, 2, 201-220.

Emiliani, C. (1993). Extinction and viruses. Biosystems, 31, 155-159.

Elsila, J. E., Glavin, D. P., Dwokin, J. P. (2009). Cometary glycine detected in samples returned by Stardust. Meteoritics & Planetary Science 44, Nr 9, 1323–1330.

Embley TM. (2006) Multiple secondary origins of the anaerobic lifestyle in eukaryotes. Philos Trans R Soc Lond B Biol Sci 361:1055–1067.

Embley TM, Martin W. (2006) Eukaryotic evolution, changes and challenges. Nature 440:623–630.

Engelstidter, J., & Hurst , G. D. D. (2007). The Impact of Male-Killing Bacteria on Host Evolutionary Processes. Genetics, 175, 245-254.

Erwin, D. H. (2001). Lessons from the past: Biotic recoveries from mass extinctions. Proceedings of the National Academy of Sciences of the USA (PNAS), 98 (10): 5399-5403.

Esser C, et al., (2004) A genome phylogeny for mitochondria among alpha-proteobacteria and a predominantly eubacterial ancestry of yeast nuclear genes. Mol Biol Evol 21:1643–1660.

Esser, C., Martin, W., & Dagan, T. (2007). The origin of mitochondria in light of a fluid prokaryotic chromosome model. Biol. Lett. 3, 180–184.

Evans, C., et al., (2009). Viral-mediated lysis of microbes and carbon release in the sub-Antarctic and Polar Frontal zones of the Australian Southern Ocean. Environmental Microbiology Reports. 11, 2924-2934.

Eyles, N., & Eyles, C.H. (1989). Glacially-influenced deep-marine sedimentation of the Late Precambrian Gaskiers Formation, Newfoundland, Canada. Sedimentology. 36, 601–620.

Fajardo A., (2008). The neglected intrinsic resistome of bacterial pathogens. PLoS ONE. 3, e1619.

Falkowski, P. G., & Godfrey, L. V. (2008). Electrons, life and the evolution of Earth's oxygen cycle Phil. Trans. R. Soc. B 27 363 no. 1504 2705-2716.

Fanning, C. M. & Link, P. K. (2004) Geology 32 , 881-884.

Farquhar, J., et al., (2000). Science, 289, 756.

Feder, M.E., and Hofmann, G. E. (1999). Heat-shock proteins, molecular chaperones, and the stress response: evolutionary and ecological physiology. Annu Rev Physiol 61:243–282.

Fedorov, A., Merican, A.F. Gilbert, W. (2002) Large-scale comparison of intron positions among animal, plant, and fungal genesPNAS December 10, 2002 vol. 99 no. 25 16128-16133.

Fedorov, A., Roy, S., Fedorova, L., Gilbert, W. (2003) Mystery of intron gain. Genome Res. 13:2236–2241.

Fekete A., et al., (2004). Simulation experiments of the effect of space environment on bacteriophage and DNA thin films. Advances in Space Research 33, 1306–1310.

Fekete, A., et al., (2005). DNA damage under simulated extraterrestrial conditions in bacteriophage T7, Advances in Space Research, 36, 303-310.

Feng D-F, Cho G, Doolittle RF. Determining divergence times with a protein clock: update and reevaluation. Proceedings of the National Academy of Sciences (USA). 1997;94:13028–13033.

Ferrara A.M. (2006) Potentially multidrug-resistant non-fermentative Gram-negative pathogens causing nosocomial pneumonia. Int. J. Antimicrob. Agents. 27, 183–195.

Filadi, R., Zampese, E., Pozzan, T., Pizzo, P., Fasolato, C., (2012) Endoplasmic Reticulum-mitochondria connections, calcium cross-talk and cell fate: a closer inspection, Endoplasmic Reticulum Stress in Health and Disease, 75-106

Firestone, R., (2009). The Case for the Younger Dryas Extraterrestrial Impact Event. Journal of Cosmology, 2, 256-285.

Flanner, B. P., Roberage, W., & Rybicki, G. B., 1980, The penetration of diffuse ultraviolet radiation into interstellar clouds. The Astrophysical Journal, 236, 598-608.

Flint, S. J., Enquist, L. W., & Racaniello, V. R. (2009). Principles of Virology. ASM Press.

Folk, R. L., Lynch, F. L. (1997). Nanobacteria are alive on Earth as well as Mars [abstract], in Proceedings of SPIE The International Society for Optical Engineering. 3111, 407-419.

Forterre, P. (2006). The origin of viruses and their possible roles in major evolutionary transitions. Virus Res. 2006 Apr;117(1):5-16.

Forterre, P., Prangishvili, D. (2013) The major role of viruses in cellular evolution: facts and hypotheses, Current opinion in virology, http://dx.doi.org/10.1016/j.coviro.2013.06.013

Fortey, R.A., et al. (1997). The Cambrian evolutionary explosion' recalibrated. BioEssays, 19, 429–34.

Fraser, C. M., et al., (1995). The Minimal Gene Complement of Mycoplasma genitalium Science, 270, 397 - 404. .

Fraser, H. J., Martin, R.S., McCoustra, D., Williams, D.A. (2002). The Molecular Universe, Astronomy and Geophysics, 43, 10.

Friedman, W.E. (2006). Embryological evidence for developmental lability during early angiosperm evolution. Nature 441: 337-340.

Friedman W.E., et al., (2004). The evolution of plant development. American Journal of Botany 91: 1726-1741.

Frost LS, Leplae R, Summers AO, Toussaint A. (2005). Mobile genetic elements: the agents of open source evolution. Nat. Rev. Microbiol. 3:722–732.

Fukue, T., et al. (2010) Extended High Circular Polarization in the Orion Massive Star Forming Region: Implications for the Origin of Homochirality in the Solar System. http://arxiv.org/pdf/1001.2608.

Garlida, K.D., et al., (2003). Mitochondrial potassium transport: the role of the mitochondrial ATP-sensitive K+ channel in cardiac function and cardioprotection Biochimica et Biophysica Acta (BBA) - Bioenergetics, 1606, 1-21.

Gehling, J.G. (1987). Earliest known echinoderm — a new Ediacaran fossil from the Pound Subgroup of South Australia. Alcheringa 11:337-345.

Gehring, W. J., (1996). The master control gene for morphogenesis and evolution of the eye. Genes Cells, 1, 11-15.

Gehring, W. J., & Ikeo, K. (1999). Pax 6: Mastering eye morphogenesis and eye evolution. Trends in Genetics. 15, 371–377.

Gehling, J. G. and Rigby, J. K., (1996). J. Palaeontol., 70, 185–195.

Gehrz, R. (1988). Sources of Stardust in the Galaxy. Journal: Interstellar Dust, 135, 445.

Geller, M.J. and Huchra, J.P. (1990). Mapping the Universe. Science, 246, 897-903.

Gerdes K., . Rasmussen P.B., Molin S. (1986). Unique type of plasmid maintenance function: postsegregational killing of plasmid-free cells. Proc. Natl Acad. Sci. USA. 83, 3116–3120.

Gibbs, C. J., et al., (1978). Unusual resistance to ionizing radiation of the viruses of kuru, Creutzfeldt-Jakob disease, and scrapie PNAS, 75, 6268-6270.

Gibson, C. H., and Wickramasinghe, N. C. (2010). The Imperatives of Cosmic Biology. Journal of Cosmology, 2010, Vol 5, 1101-1120.

Gibson, G., & Hogness, D. S. (1996). Effects of polymorphism in the Drosophilia regulatory gene Ultrabithorax on homoetic stability. Science 271, 200-203. Gilbert, W. (1978) Why genes in pieces? Nature 271:501.

Gilbert, L. I. (2009) Insect Development: Morphogenesis, Molting and Metamorphosis. Academic Press.

Gilbert, W. (1986). The RNA world. Nature 319, 618.

Gilbert, W. (1987)The exon theory of genes. Cold Spring Harbor Symp. Quant. Biol. 52, 901–905.

Gilichinsky, D. A. (2002a). Permafrost Model of Extraterrestrial Habitat. In G. Horneck & C. Baumstark-Khan. Astrobiology, Springer.

Gilichinsky, D. (2002b) in Encyclopedia of Environmental Microbiology, ed. Bitton, G. (Wiley, New York), pp. 2367-2385.

Gilli, G. and Gilli. P. (2009). The Nature of the Hydrogen Bond: Oxford University Press.

Gilliver M.A., et al., (1999). Antibiotic resistance found in wild rodents. Nature. 401, 233–234.

Gillon, M., Pont, F., Demory, B.-O. et al. (2007), A&A, 472, L13.

Gingerich, P. D. (2001) Rates of evolution on the time scale of the evolutionary process. Genetica 112-113, 127–144

Goff, P. H., Gao, Q., & Palese, P. (2012) A Majority of Infectious Newcastle Disease Virus Particles Contain a Single Genome, while a Minority Contain Multiple Genomes, Journal of Virology, 86, 10852-10856.

Gladman, J. E.,. et al. (1996). The exchange of impact ejecta between terrestrial Planets, Science. 271, 1387-1392.

Glickson, A. (2009). Mass extinction of species: The role of external forcing. Journal of Cosmology, 2

Goertzel, B. and Combs, A. (2010). Water Worlds, Naive Physics, Intelligent Life, and Alien Minds. Journal of Cosmology, 5, 897-904.

Gong, Y, M., Xu, R., & Hu, B. (2008). Endolithic fungi: A possible killer for the mass extinction of Cretaceous dinosaurs Science in China. Series D: Earth Sciences, 51, 801-807.

Glavin D. P., et al., (2006). liquid chromatography-time of flight-mass spectrometry. Meteoritics & Planetary Science 41(6):889–902.

Gogarten JP, Doolittle WF, Lawrence JG. (2002). Prokaryotic evolution in light of gene transfer. Mol. Biol. Evol. 19, 2226–2238.

Gogarten JP, Townsend JP. (2005). Horizontal gene transfer, genome innovation and evolution. Nat. Rev. Microbiol. 3:679–687.

Gold T (1992). The deep, hot biosphere. Proc. Natl. Acad. Sci. U.S.A. 89, 6045–6049.

Gott, J. R., et al., Astrophys. J., 624, 463 (2005).

Gould, S. J., (2001). The Structure of Evolutionary Theory. Belknap Press of Harvard University Press.

Gray, M. W. et al., (1999). Mitochondrial Evolution Science, 283, 1476 - 1481

Grenet K., et al., (2004). Antibacterial resistance, Wayampis Amerindians, French Guyana. Emerg. Infect. Dis. 10, 1150–1153.

Gu, X. (1998). Early Metazoan Divergence Was About 830 Million Years Ago. J. Mol. Evol. 47, 369-371.

Guetg, C. et al. (2012). Inheritance of Silent rDNA Chromatin Is Mediated by PARP1 via Noncoding RNA, Molecular Cell, 45 790–800

Guillermo G. G., Brownleea, D., and Ward, P. (2001). The Galactic Habitable Zone: Galactic Chemical Evolution. Icarus, 152, 185-200.

Hacker, J., Kaper, J.B. (2000). Pathogenicity islands and the evolution of microbes. Annu.

Rev. Microbiol. 54:641–679.

Hadrys T., DeSalle, R., Sagasser, S. Fischer,N., Schierwater, B., (2005). The Trichoplax PaxB Gene: A Putative Proto-PaxA/B/C Gene Predating the Origin of Nerve and Sensory Cells. Molecular Biology and Evolution, 22, 1569-1578.

Harland, W. B. (2007). Origins and assessment of snowball Earth hypotheses --Geological Magazine, 144, 633-642.

Haldane, J. B. (2009) What I Require From Life: Writings on Science and Life From J.B.S. Haldane Oxford University Press, USA.

Hall, J. A., et al., (2003). The search for viruses through the fossil record. Goldschmidt Conference Abstracts 2003 A129.

Hansen, C. J. et al., (2004) Stellar Interiors - Physical Principles, Structure, and Evolution. Springer.

Häring, M., et al., (2005). Viral Diversity in Hot Springs of Pozzuoli, Italy, and Characterization of a Unique Archaeal Virus, Acidianus Bottle-Shaped Virus, from a New Family, the Ampullaviridae Journal of Virology, 79, 9904-9911.

Harris, J. K., et al., (2003). The Genetic Core of the Universal Ancestor Genome Res. 13: 407-412.

Harris, M. J., Wickramasinghe, N.C., Lloyd, D. et al., (2002). Detection of living cells in stratospheric samples. Proc. SPIE, 4495, 192-198.

Hartmann, L., Heitsch, F., and Ballesteros-Paredes, J. (2009). Dynamic star formation. Rev Mex A A (Serie de Conferencias), 35, 66

Hautmann, M. (2004). Effect of end-Triassic CO_2 maximum on carbonate sedimentation and marine mass extinction. Facies Volume 50, 1-13.

Hayes F. (2003). Toxins-antitoxins: plasmid maintenance, programmed cell death, and cell cycle arrest. Science. 301, 1496–1499.

Hedges, S.B. et al. (2001) BMC Evolutionary Biology, 1 : 4-14.

Hedges, S,B, Blair, J,E,, Venturim M,L,, Shoem J.L., (2004). A molecular timescale of eukaryote evolution and the rise of complex multicellular life. BMC Evol Biol.28, 2.

Hedges SB. (2002). The origin and evolution of model organisms. Nature Reviews Genetics. ;3:838–849. doi: 10.1038/nrg929.

Hedges SB, Blair JE, Venturi ML, Shoe JL., (2004). A molecular timescale of eukaryote evolution and the rise of complex multicellular life. BMC Evol Biol. Jan 28;4:2.

Hendrix. R. W. (2004). Hot new virus, deep connections PNAS 101, 7495-7496.

Henze K, Badr A, Wettern M, Cerff R, Martin W (1995) A nuclear gene of eubacterial origin in Euglena gracilis reflects cryptic endosymbioses during protist evolution Proc Natl Acad Sci USA 92:9122–9126.

Herbst, E., Klemperer, W. 1973. The formation and depletion of molecules in dense interstellar clouds. Astrophysical Journal, 185, 505-533.

Herrmann, J.M., & Neupert, W. (2000). Protein transport into mitochondria. Curr Opin Microbiol, 3, 210–214.

Herrero, A., Flores, E.,(2008). The Cyanobacteria: Molecular Biology, Genomics and Evolution Caister Academic Press.

Hijnen,W.A.M. et al., (2006). Inactivation credit of UV radiation for Viruses, Bacteria and protozoan (oo)cysts in water: A review. Water Research, 40, 3-22.

Hiraga S., Jaffe A., Ogura T., . Mori H., Takahashi H. (1986). F plasmid ccd mechanism in Escherichia coli. J. Bacteriol. 166, 100–104.

Hoffmann, K. H., Condon, D. J., Bowring, S. A. & Crowley, J. L. (2004) Geology 32 , 817-820.

Hoffman, P. F., Kaufman, A. J., Halverson, G. P. and Schrag, D. P. (1998). A Neoproterozoic snowball Earth. Science 281, 1342-1346.

Holland H.D (2006) The oxygenation of the atmosphere and oceans. Phil. Trans. R. Soc. B.

361, 903–915.

Hoover, R. B. (1997). Meteorites, Microfossils, and Exobiology" [abstract] in Instruments, Methods, and Missions for the Investigation of Extraterrestrial Microorganisms. In Hoover, R. B., Editor, Proceedings of SPIE Vol. 3111, 115-136.

Hoover, R.B., (1998). Meteorites, Microfossils, and Exobiology" [abstract] in Instruments, Methods, and Missions for the Investigation of Extraterrestrial Microorganisms. In Hoover, R. B. Editor, Proceedings of SPIE Vol. 3111, 115-136.

Hoover, R.B. (2005). In R.B. Hoover, A.Y. Rozanov and R.R. Paepe (eds), Perspectives in Astrobiology, Amsterdam, IOS Press, 366, 43.

Hoover R.B. (2006). Microfossils in carbonaceous meteorites. In Cosmic Dust and Panspermia. Progress towards unravelling our cosmic ancestry, International Conference at Cardiff University, Wales, 5-8 Sept. 2006.

Hoover, R.B. (2011). Fossils of Cyanobacteria in CI1 Carbonaceous Meteorites, Journal of Cosmology, 13.

Hoover, R. B., Rozanov, A., (2003). Microfossils, biominerals and chemical biomarkers in Meteorites, in: Instruments Methods and Missions for Astrobiology VI, edited by: Hoover, R. B., Rozanov, A. Yu., and Lipps, J. H., Proc. SPIE 4939, 10-27.

Horgan, J. (1991). In the beginning. Scientific American, 264, 116-125.

Horiike T, Hamada K, Miyata D, Shinozawa T. (2004)The origin of eukaryotes is suggested as the symbiosis of pyrococcus into gamma-proteobacteria by phylogenetic tree based on gene content. J Mol Evol 59:606–619.

Horneck, G. (1993). Responses ofBacillus subtilis spores to space environment: Results from experiments in space Origins of Life and Evolution of Biospheres 23, 37-52.

Horneck, G., Bücker, H., Reitz, G. (1994). Long-term survival of bacterial spores in space. Advances in Space Research, Volume 14, 41-45.

Horneck, G., Eschweiler, U., Reitz, G., Wehner, J., Willimek, R., Strauch, G. (1995). Biological responses to space: results of the experiment "Exobiological Unit" of ERA on EURECA I. Advances in Space Research 16, 105-118.

Horneck, G., Stöffler, D., Eschweiler, U., Hornemann, U. (2001). Bacterial spores survive simulated meteorite impact Icarus 149, 285.

Horneck, G., Rettberg, P., Reitz, G., Wehner, J., Eschweiler, U., Strauch, K., Panitz, C., Starke, V., Baumstark-Khan, C. (2001). Origins of Life and Evolution of Biospheres 31, 527-547.

Horneck, G. Mileikowsky, C., Melosh, H. J., Wilson, J. W. Cucinotta F. A., Gladman, B. (2002). Viable Transfer of Microorganisms in the solar system and beyond, In G. Horneck & C. Baumstark-Khan. Astrobiology, Springer.

Hotopp JC, Clark ME, Oliveira DC, Foster JM, Fischer P, Torres MC, Giebel JD, Kumar N, Ishmael N, Wang S, et al. (2007). Widespread lateral gene transfer from intracellular bacteria to multicellular eukaryotes. Science, 317:1753–1756

Howe, C. J. et al., (2008). The origin of plastids Phil. Trans. R. Soc. B 27, . 363, 2675-2685.

Hoyle, F. (1974) Intelligent Universe.

Hoyle, F., (1977). Polysaccharides and the infrared spectra of galactic sources", Nature, 268, 610.

Hoyle, F. (1978). Lifecloud - The Origin of Life in the Universe, J.M. Dent and Sons.

Hoyle, F. and Wickramasinghe, N.C., 2000. Astronomical Origins of Life: Steps towards Panspermia. Kluwer Academic Press.

Huder, J. B., et al. (2002). Identification and characterization of two closely related unclassifiable endogenous retroviruses in pythons (Python molurus and Python curtus). J. Virol. 76:7607-7615.

Huff, E. M., and Stahler, S. W. (2006). Star formation in space and time: The Orion Nebula cluster. The Astrophysical journal 644, 355-363,

Alien Viruses, Genetics, Cambrian Explosion: Humans

Hughes JF, Coffin JM. (2001). Evidence for genomic rearrangements mediated by human endogenous retroviruses during primate evolution. Nat Genet ;29:487 -489.

Hyde, W. T., Crowley, T. J., Baum, S. K. and Peltier, W. R. (2000). Neoproterozoic 'snow-ball earth' simulations with a coupled climate/ice-sheet model. Nature 405, 425-429.

IHGSC (2001). International Human Genome Sequencing Consortium. 2001. Initial sequencing and analysis of the human genome. Nature 409:860–921.

Ingles-Prieto, et al. (2013) Conservation of Protein structure over Four Billion years. Structure, 21, 1690-1697

Iyer LM, Koonin EV, Aravind L. (2004). Evolution of bacterial RNA polymerase: implications for large-scale bacterial phylogeny, domain accretion, and horizontal gene transfer. Gene, 335:73–88.

Jacobsen, S. B., (2005). The Hf-W system and the origin of the Earth and Moon. Annual Review of Earth and Planetary Sciences. 33, 531-570.

Jaffe A., Ogura T., Hiraga S. (1985). Effects of the ccd function of the F plasmid on bacterial growth. J. Bacteriol. 163, 841–849.

Jagoutz, E., Sorowka, A., Vogel, J. D., Wnke, H. (1994). ALH 84001: Alien or progenitor of the SNC family? Meteoritics, 29, 478-479.

Jain, R., et al., (1999). Horizontal gene transfer among genomes: The complexity hypothesis, PNAS March 30, vol. 96 no. 7 3801-3806.

Joachimski, M. J., & Buggisch, W. (2002). Conodont apatite $\delta18O$ signatures indicate climatic cooling as a trigger of the Late Devonian mass extinction. Geology, 30, 711-714.

Jiggins, F. M. (2003). Male-Killing Wolbachia and Mitochondrial DNA: Selective Sweeps, Hybrid Introgression and Parasite Population Dynamics. Genetics, 164, 5-12.

John, B., & Miklos, G. (1988). The Eucaryotic Genome in Development and Evolution. Allen & Unwin, London.

Johnson W.E and Coffin, J.M (1999). Constructing primate phylogenies from ancient retrovirus sequences, Proc Natl Acad Sci USA 96 10254–10260.

Jones, M.J., et al., (2004). An Introduction to Galaxies and Cosmology. Cambridge University Press. pp. 50-51.

Jones, A. R. (2009). The next mass extinction: Human evolution or human eradication. Journal of Cosmology, 2009, 2, 316-333

Jose, M. V. et al., (2010). How Universal is the Universal Genetic Code? A Question of ExtraTerrestrial Origins. Journal of Cosmology, 5,

Jordan, K., et al., (2003). Origin of a substantial fraction of human regulatory sequences from transposable elements Trends in Genetics, 19, 68-72.

Jordan, T. H. (1979). Structural Geology of the Earth's Interior". Proceedings National Academy of Science, 76, 4192–4200.

Joseph, R. (1979). Effects of rearing environment and sex on learning, memory, and competitive exploration. Journal of Psychology, 101, 37-43.

Joseph, R. & Casagrande, V. A. (1978). Visual field defects and recovery following lid closure in a prosimian primate. Behavioral Brain Research, 1, 150-178.

Joseph, R., & Casagrande, V. A. (1978). Visual field defects and morphological changes resulting from monocular deprivation in primates. Proceedings of the Society for Neuroscience, 4, 1982, 2021.

Joseph, R., & Gallagher, R. E. (1980). Gender and early environmental influences on learning, memory, activity, overresponsiveness, and exploration. journal of Developmental Psychobiology, 13, 527-544.

Joseph, R. (1997). Life on Earth Came From Other Planets, University Press.

Joseph, R. (1999). Environmental influences on neural plasticity, the limbic system, and emotional development and attachment, Child Psychiatry and Human Development. 29, 187-203.

Joseph, R. (2000a). Astrobiology, the origin of life, and the Death of Darwinism. University Press, California.

Joseph, R. (2000b). Fetal brain behavioral cognitive development. Developmental Review, 20, 81-98.

Joseph, R. (2009a). Life on Earth came from other planets. Journal of Cosmology, 1, 1- 56.

Joseph, R. (2009b). The evolution of life from other planets. The first Earthlings. Interplanetary genetic messengers. Extraterrestrial horizontal gene transfer. The genetics of eukaryogenesis and mitochondria metamorphosis, Journal of Cosmology, 1, 100-150.

Joseph, R. (2009c). Genetics and Evolution of Life From Other Planets: Viruses, Bacteria, Archae, Eukaryotes..., Journal of Cosmology, 2009, 1, 151-200.

Joseph, R. (2009d). The Evolution of Life From Other Planets: Part 4. Genes, Microbes, Metazoan Metamorphosis Cambrian Explosion and the Genetically Engineered Earth, Journal of Cosmology, 2009, 1, 200-250.

Joseph, R. (2009e). Extinction, Metamorphosis, Evolutionary Apoptosis, and Genetically Programmed Species Mass Death, Journal of Cosmology, 2009, 2, 235-255.

Joseph (2010a) Climate change: The first four billion years. The biological cosmology of global warming and global freezing. Journal of Cosmology, 2010, 8, 2000-2020.

Joseph, R.(2010b).The Origin of Eukaryotes: Archae, Bacteria, Viruses and Horizontal Gene Transfer , In "Abiogenesis and the Origins of Life" Edited by M. J. Russell, Cosmology Science Publishers. Cambridge, MA

Joseph, R., Schild, R. (2010a). Biological cosmology and the origins of life in the universe. Journal of Cosmology, 5, 1040-1090.

Joseph, R. & Schild, R. (2010b). Origins, Evolution, and Distribution of Life in the Cosmos: Panspermia, Genetics, Microbes, and Viral Visitors From the Stars. In "The Biological Big Bang," Edited by Chandra Wickramasinghe, Science Publishers, Cambridge, MA

Joseph, R. Wickramasinghe, N. C. (2010). Comets and contagion: Evolution and diseases srom space. Journal of Cosmology, 2010, 7, 1750-1770.

Joseph, R. & Wickramasinghe, N. C., (2011).Genetics Indicates Extraterrestrial Origins for Life: The First Gene Journal of Cosmology, 2011, Vol. 16.

Joseph, R. (2012a). Evidence for Extraterrestrial Extremophiles and Plasmas in the Thermosphere Cosmology, 3/9/2012, 23-52.

Joseph, R. (2012b). Neuroscience. University Press. California.

Jung, P-M., et al., (2009). Radiation sensitivity of polioVirus, a model for noroVirus, inoculated in oyster (Crassostrea gigas) and culture broth under different conditions. Radiation Physics and Chemistry, 8, 597-599.

Jura, M., Bohac, C.J., Sargent, F., Forrest, B.W.J., Green, J.D., Watson, D.M., Sloan, G.C., Keller,L.D., Markwick-Kemper, F. Chen, C.H., and Najita, J. (2005). Polycyclic aromatic hydrocarbons orbiting HD233517, and evolved oxygen rich red giant, Astrophys. J. (Letters) 637, L45.

Kado CI. (1998). Origin and evolution of plasmids. Antonie Van Leeuwenhoek, 73:117–126.

Kalirai, J. S., Bergeron, P., Hansen, B. M. S., Kelson, D. D., Reitzel, D. B., Rich, R.M., Richer, H. B. (2007). Stellar Evolution in NGC 6791: Mass Loss on the Red Giant Branch and the Formation of Low-Mass White Dwarfs. Astrophysical Journal 671 748-760.

Kapitonov VV, Jurka J (2005) RAG1 core and V(D)J recombination signal sequences were derived from Transib transposons. PLoS Biol 3: e181.

Karner MB, DeLong EF, Karl DM (2001). Archaeal dominance in the mesopelagic zone of the Pacific Ocean". Nature 409 (6819): 507–10.

Kasting J.F, Ackerman T.P (1986). Climatic consequences of very high CO_2 levels in the earth's early atmosphere. Science. 234, 1383–1385.

Kasting, J.F., & Ono, S. (2006). Palaeoclimates: the first two billion years. Phil. Trans. R Soc. B. 361, 917–929.

Kasting, J.F., & Ono, S. (2006). Palaeoclimates: the first two billion years. Phil. Trans. R

Soc. B. 361, 917–929.

Kasting J.F, Siefert J.L (2002). Life and the evolution of Earth's atmosphere. Science. 296, 1066–1068.

Kenmochi N, Suzuki T, Uechi T, Magoori M, Kuniba M, et al. (2001) The human mitochondrial ribosomal protein genes: Mapping of 54 genes to the chromosomes and implications for human disorders. Genomics 77: 65–70.

Kensei K., et al., (2008). Formation of amino acid precursors with large molecular weight in dense clouds and their relevance to origins of bio-homochirality. Proceedings of the International Astronomical Union, 4:465-472.

Kidder, D. L., & Worsley, T. R. (2004). Causes and consequences of extreme Permo-Triassic warming to globally equable climate and relation to the Permo-Triassic extinction and recovery. Palaeogeography, Palaeoclimatology, Palaeoecology, 203, 207-237.

Kimura, H., J.-I. Ishibashi, H. Masuda, K. Kato, and S. Hanada (2007). Selective Phylogenetic Analysis Targeting 16S rRNA Genes of Hyperthermophilic Archaea in the Deep-Subsurface Hot Biosphere Appl. Environ. Microbiol. 73:2110-2117.

Kimura, H., M. Sugihara, K. Kato, and S. Hanada (2006). Selective Phylogenetic Analysis Targeted at 16S rRNA Genes of Thermophiles and Hyperthermophiles in Deep-Subsurface Geothermal Environments Appl. Environ. Microbiol. 72:21-27.

Kirshvink, J.L, Gaidos, E.J, Bertaini, L.E, Beukes, N.J, Gutzmer, J, Maepa, L.N, Steinberger, R.E. (2000). Paleoproterozoic snowball Earth: extreme climatic and geochemical global change and its biological consequences. Proceedings of the National Academy of Sciences of the United States of America, 97, 1400-1405.

Klamer, I.J., Ekers, R.D., Sadler, E.M., Hunstead, R.W. (2004), ApJ 612, L100.

Klyce, B. (2000). Microorganisms from the Moon. http://www.panspermia.org/zhmur2.htm.

Koninga, N., et al., (2008). Organic molecules in the spectral line survey of Orion KL with the Odin Satellite from 486-492 GHz and 541-577 GHz. Proceedings of the International Astronomical Union, 4, 29-30.

Koonin, EV. (2003) Comparative genomics, minimal gene-sets and the last universal common ancestor. Nature Rev. Microbiol. 1:127–136.

Koonin, E.V., et al. (2004). A comprehensive evolutionary classification of proteins encoded in complete eukaryotic genomes. Genome Biol. 5, R7.

Koonin EV. (2006). The origin of introns and their role in eukaryogenesis: a compromise solution to the introns-early versus introns-late debate? Biol Direct. Aug 14;1:22.

Koonin, E. V., (2009a). Darwinian evolution in the light of genomics. Nucleic Acids Research 37(4):1011-1034 Koonin EV. (2009b). Evolution of genome architecture. Int. J. Biochem. Cell Biol. 41:298–306

Koonin, E.V., & Wolf, Y.I. (2008). Genomics of bacteria and archaea: the emerging generalizations after 13 years. Nucleic Acids Res. 36, 6688–6719.

Kunin V, Ouzounis CA. (2003) The balance of driving forces during genome evolution in prokaryotes. Genome Res. 13:1589–1594.

Kuppers, B.O. (1990). Information and the origin of life. Cambridge, MA: MIT Press.

Kurland, CG, Collins LJ, Penny D. (2006). Genomics and the irreducible nature of eukaryote cells. Science 312:1011–1014.

Lake JA. (1988). Origin of the eukaryotic nucleus determined by rate-invariant analysis of rRNA sequences. Nature, 331:184–186.

Lake JA. (1998). Optimally recovering rate variation information from genomes and sequences: pattern filtering. Mol Biol Evol. 15:1224–1231.

Lake JA, Henderson E, Oakes M, Clark MW. (1984). Eocytes: a new ribosome structure indicates a kingdom with a close relationship to eukaryotes. Proc Natl Acad Sci USA. 81:3786–3790.

Lake JA, Rivera MC. (1994) Was the nucleus the first endosymbiont? Proc Natl Acad Sci

USA. 91:2880–2881.

Lander, E.S. et al., (2001). Human Genome Initial sequencing and analysis of the human genome Nature 409, 860-921.

Lane, N, et al., (2010). How did LUCA make a living? Chemiosmosis in the origin of life. BioEssays.

Lantzy, R., Dacey, M. and Mckenzie, F. (1977). Catastrophe theory: Application to the Permian mass extinction. Geology, 5, 724-728.

Lavie, L., et al., (2004). Human endogenous retrovirus family HERV-K(HML-5): Status, evolution, and reconstruction of an ancient eetaretrovirus in the human genome, Journal of Virology, 78, 8788-8798.

Leff, S. E., Rosenfeld, M. G. & Evans, R. M. (1986) Annu. Rev. Biochem. 55, 1091-1118.

Leipe DD, Aravind L, Koonin EV. (1999). Did DNA replication evolve twice independently? Nucleic Acids Res. 27:3389–3401.

Leininger S, Urich T, Schloter M, et al. (2006). Archaea predominate among ammonia-oxidizing prokaryotes in soils. Nature 442 (7104): 806–809.

Leister, D. (2003). Chloroplast research in the genomic age, Trends in Genetics 19, 147-56.

Lemaître, G. (1927). Un Univers homogne de masse constante et de rayon croissant rendant compte de la vitesse radiale des nbuleuses extra-galactiques, Annales de la SociScientifique de Bruxelles, 47, 49. (A homogeneous universe of constant mass and increasing radius accounting for the radial velocity of extra-galactic nebulae, Annals of the Scientific Society of Brussels, 47, 49).

Lemaître, G. (1931a) "Expansion of the universe, The expanding universe", Monthly Notices of the Royal Astronomical Society, 91, 490-501 (Expansion of the universe, The Expanding Universe, Monthly Notices of the Royal Astronomical Society, Vol. 91, p.490-501, 03/1931).

Lemaître, G. (1931b). The Beginning of the World from the Point of View of Quantum Theory, Nature 127, n. 3210, 706.

Leng-feng, Y. (1976). Mawangdui Boshu Laozi Shitan. Taipei.

Lerner, E.J. (1991). The Big Bang Never Happened, Random House, New York.

Levin, G. V. (2010). Extant Life on Mars: Resolving the Issues. Journal of Cosmology, 5, 920-929.

Li, C-W, Chen, J-Y, Hua. T-E. (1998). Precambrian Sponges with Cellular Structures. Science, 279, 879-882.

Li, Z. et al. (2003). Geochronology of Neoproterozoic synrift magmatism in the Yangtze Craton, South China and correlations with other continents: evidence for a mantle superplume that broke up Rodinia. Precambrian Res. 122, 85–109.

Liebert, J., Arnett, E., Holberg, J., Williams, K. (2005a). Sirius. Astrophysical Journal Letters. 630, L69-L72.

Liebert, K., Young, P. A., Arnett, E., Holberg, J. B., Williams, K. A. (2005b) The Age and Progenitor Mass of Sirius B. The Astrophysical Journal Letters 630, L69-L72.

Lindell, D., et al., (2004). Transfer of photosynthesis genes to and from Prochlorococcus viruses. Proc Natl Acad Sci, 101, 11013-11018.

Lindell, D., et al., (2005). Photosynthesis genes in marine viruses yield proteins during host infection. Nature, 438, 86-89.

Lipps, G., (2006). Plasmids and viruses of the thermoacidophilic crenarchaeote Sulfolobus, Extremophiles, 10, 17-28.

Livermore D.M., et al., (2001). Antibiotic resistance in bacteria from magpies (Pica pica) and rabbits (Oryctolagus cuniculus) from West Wales. Environ. Microbiol. 3, 658–661.

LogsdonJ. M. (1998). The recent origins of spliceosomal introns revisited Curr. Opin. Genet. Dev. 8 , 637-648.

Lonergan, D. J., Jenter, H. L., Coates, J. D., Phillips, E. J., Schmidt, T. M. & Lovley, D. R. (1996). Phylogenetic analysis of dissimilatory Fe(III)-reducing bacteria.J Bacteriol 178, 2402-

2408.

López-Sánchez, P., Costas,J. C., and Naveir, H. F. (2005). Paleogenomic Record of the Extinction of Human Endogenous Retrovirus ERV9. Journal of Virology, 79, 6997-7004.

Lorenc, A., and Makalowski, W. (2003). Transposable elements and vertebrate protein diversity. Genetica 118:183-191.

Lovett, S. T. (2006). Microbiology: Resurrecting a broken genome. Nature 443, 517-519.

Lovett S.T., Hurley R.L., Sutera V.A., Jr, Aubuchon R.H., . Lebedeva M.A. (2002). Crossing over between regions of limited homology in Escherichia coli. RecA-dependent and RecA-independent pathways. Genetics. 160, 851–859.

Lovelock, J. E., & Margoulis, L. (1974). Atmospheric homeostasis by and for the biosphere - the Gaia hypothesis. Tellus, 26, 2-10.

Lovley, D. R. (1991). Dissimilatory Fe(III) and Mn(IV) reduction.Microbiol Rev 55, 259-287.

Lupski, J. R. (2013) Genome Mosaicism—One Human, Multiple Genomes, Science, 341, 358-359.

Lynch M. (2007). The Origins of Genome Architecture, Sunderland, MA: Sinauer Associates.

Lynch, M,, & Conery, J.S. (2000). The evolutionary fate and consequences of duplicate genes. Science. 290, 1151–1155.

Lynch, M., O'Hely, M., Walsh, B., & Force, A. (2001). The probability of preservation of a newly arisen gene duplicate. Genetics. 159, 1789–1804.

MacLeod, N. (2000). Extinction! First Science.com.

MacLeod, N. (2001). Extinction. in Encyclopaedia of Life Sciences. Macmillan, London.

Maeder DL, Anderson I, Brettin TS, Bruce DC, Gilna P, Han CS, Lapidus A, Metcalf WW, Saunders E, Tapia R, et al. (2006). The Methanosarcina barkeri genome: comparative analysis with Methanosarcina acetivorans and Methanosarcina mazei reveals extensive rearrangement within methanosarcinal genomes. J. Bacteriol. 188:7922–7931.

Makarova, K. S., et al., (2005). Ancestral paralogs and pseudoparalogs and their role in the emergence of the eukaryotic cell Nucleic Acids Research, 33(14):4626-4638.

Mandel MA, Yanofsky MF. (1995). A gene triggering flower formation in Arabidopsis. Nature. 377, 482-483.

Mannella, C.A. (2006). Structure and dynamics of the mitochondrial inner membrane cristae . 1763, 542–548.

Manning, C. E., Mojzsis, S. J , Harrison, T. M. (2006). Geology, age and origin, of supracrustal rocks at Akilia, West Greenland. American Journal of Science, 306, 303-366.

Marcaide, J. M., and Weiler, K. W. (2005) Cosmic Explosions. Springer.

Marquis, R. E., and S. Y. Shin. (1994). Mineralization and responses of bacterial spores to heat and oxidative agents. FEMS Microbiol. Rev. 14:375-380.

Margulis, L. (1988). Symbiotic planet. In Basic Books 1998 New York, NY:Basic Books.

Margulis, L, et al. (1997). Microcosmos; Four Billion Years of Evolution from Our Microbial Ancestors, University of California Press.

Margulis, L., Sagan, D., & Thomas, L. (1997). Microcosmos; Four Billion Years of Evolution from Our Microbial Ancestors, University of California Press.

Marquis RE, Sim J, Shin SY. (2006). Molecular mechanisms of resistance to heat. J Appl Microbiol. 101(3):514-25.

Martel, J., Young, J D-E. (2008). Purported nanobacteria in human blood as calcium carbonate nanoparticles PNAS, 105 5549-5554

Martin W, Koonin EV. (2006). Introns and the origin of nucleus-cytosol compartmentation. Nature 440:41–45.

Martin W, Muller M. (1998). The hydrogen hypothesis for the first eukaryote. Nature 392:37–41.

Martin W, Rujan T, Richly E, Hansen A, Cornelsen S, Lins T, Leister D, Stoebe B, Hasegawa M, Penny D. (2002). Evolutionary analysis of Arabidopsis, cyanobacterial, and chloroplast genomes reveals plastid phylogeny and thousands of cyanobacterial genes in the nucleus. Proc. Natl Acad. Sci. USA 99:12246–12251.

Martinez, J. L. (2009). The role of natural environments in the evolution of resistance traits in pathogenic bacteria Proc. R. Soc. B, 276, 2521-2530.

Martinez J.L., Perez-Diaz J.C. (1990). Journal of Antimicrobial Chemotherapy, 26, 301-305.

Martinez J.L., Perez-Diaz J.C. (1990) Cloning of the determinants for microcin D93 production and analysis of three different D-type microcin plasmids. Plasmid. 23, 216–225.

Martinez J.L., . Baquero F., Andersson D.I. (2007). Predicting antibiotic resistance. Nat. Rev. Microbiol. 5, 958–965.

Maruyama, S., et al., (2011). Eukaryote-to-eukaryote gene transfer gives rise to genome mosaicism in euglenids, BMC Evolutionary Biology, 11, 105.

Marzoli, L. M. et al. 1999, Science 284. Extensive 200-million-year-old continental flood basalts of the Central Atlantic Magmatic Province, pp. 618-620.

Mastrapaa, R.M.E., Glanzbergb, H ., Headc, J.N., Melosha, H.J, Nicholson, W.L. (2001). Survival of bacteria exposed to extreme acceleration: implications for panspermia, Earth and Planetary Science Letters 189, 30 1-8.

Mautner, M. (2010). Seeding the Universe with Life: Securing Our Cosmological Future. Journal of Cosmology, 5

Mayer, J. Meese, E. (2005). Human endogenous retroviruses in the primate lineage and their influence on host genomes. Cytogenetic and Genome Reserach, 110, 1-4.

Mayer, J., E. Meese, and N. Mueller-Lantzsch. (1998). Human endogenous retrovirus K homologous sequences and their coding capacity in Old World primates. J. Virol. 72:1870-1875.

Mazel D. (2006). Integrons: agents of bacterial evolution. Nat. Rev. Microbiol. 4, 608–620.

McCarthy, P.J., Van Breugel, W., Spinrad, H., Djorgovski, S. (1987), ApJ 321, L29.

McLysaght, A., Hokamp, K., & Wolfe, K.H., (2002). Extensive genomic duplication during early chordate evolution. Nat Genet. 31, 28-9.

McKay, D. S., Gibson Jr., E. K., Thomas-Keprta, K.L., Vali, H., Romanek, C. S., Clemett, S. J., Chillier, X.D. F., Maechling, C. R., Zare, R. N. (1996). Search for Past Life on Mars: Possible Relic Biogenic Activity in Martian Meteorite ALH84001. Science 273 (5277): 924-930.

McClintock, J. E. (2004). Black hole. World Book Online Reference Center. World Book, Inc.

McKee, J. K. (2009). Contemporary mass extinction and the human population imperative. Journal of Cosmology, 2009, 2, 300-308

Medstrand P, Mager DL. (1998). Human-specific integrations of the HERV-K endogenous retrovirus family. J Virol. 72(12):9782-7.

Medstrand, P., D. et al. (1997). Structure and genomic organization of a novel human endogenous retrovirus family: HERV-K (HML-6). J. Gen. Virol. 78:1731-1744.

Medstrand P, van de Lagemaat LN, Mager DL. (2002). Retroelement distributions in the human genome: variations associated with age and proximity to genes. Genome Res12:1483 -1495.

Melott, A. L., et al. (2004). Did a gamma-ray burst initiate the late Ordovician mass extinction? International Journal of Astrobiology, 3, 55-61.

Mentel, M. Martin, W. (2008) Energy metabolism among eukaryotic anaerobes in light of Proterozoic ocean chemistry Phil. Trans. R. Soc. B 27 363 no. 1504 2717-2729.

Mi S, Lee X, Li X, et al. (2000). Syncytin is a captive retroviral envelope protein involved in human placental morphogenesis. Nature; 403:785 -789.

Miller, W. J., et al., (1999). Molecular domestication—more than a sporadic episode in evolution. Genetica 107:197-207.

Miller, V. M. et al., (2004). Evidence of nanobacterial-like structures in calcified human arteries and cardiac valves. Am J Physiol Heart Circ Physiol 287: H1115-H1124.

Mills, R. E. et al. (2011). Mapping copy number variation by population scale genome sequencing, Nature, 470, 59-65.

Milner-White, E.J., Russell, M.J. (2010). Polyphosphate synergy and the organic takeover at the origin of life. Journal of Cosmology, 5, 3217-3229.

Mirkin BG, Fenner TI, Galperin MY, Koonin EV. (2003). Algorithms for computing parsimonious evolutionary scenarios for genome evolution, the last universal common ancestor and dominance of horizontal gene transfer in the evolution of prokaryotes. BMC Evol. Biol. 3:2.

Mitchell, F. J., & Ellis, W. L. (1971). Surveyor III: Bacterium isolated from lunar retrieved TV camera. In A.A. Levinson (ed.). Proceedings of the second lunar science conference. MIT press, Cambridge.

Miyahara, H., Yokoyama, Y., Masuda, K. (2008b). Possible link between multi-decadal climate cycles and periodic reversals of solar magnetic field polarity, Earth and Planetary Science Letters, 272, 290-295.

Miyahara, H., Yokoyama, Y., Yamaguchi, Y. T. (2009). Influence of the Schwabe/Hale solar cycles on climate change during the Maunder Minimum. In: Kosvichev, A. G., Andrei, A. H., Rozelot, J.-P. (Eds.), Proceedings of IAU symposium No. 264, 427-433.

Molina, E., Arenillas, I. and Arz, J. A. (1996). The Cretaceous/Tertiary boundary mass extinction in planktic foraminifera at Agost, Spain. Rev. Micropaléont, 39 (3): 225-243.

Mooi, R., & Bruno, D. (1999). Evolution within a bizarre phylum: Homologies of the first echinoderms. American Zoologist, 38, 965–974.

Mojzsis, S.J., Arrhenius, G., McKeegan, K.D., Harrison, T.M., Nutman, A.P., Friend, C.R.L. (1996). Evidence for life on Earth before 3,800 million years ago. Nature 384, 55–59.

Moritz E.M., . Hergenrother P.J. (2007). Toxin-antitoxin systems are ubiquitous and plasmid-encoded in vancomycin-resistant enterococci. Proc. Natl Acad. Sci. USA. 104, 311–316.

Moser, D. P. et al., 2005. Desulfotomaculum and Methanobacterium spp. Dominate a 4- to 5-Kilometer-Deep Fault. Applied and Environmental Microbiology, 71, 8773-8783.

Moss, B. O., Elroy-Stein, Mizukami, T., et al. (1990). New mammalian expression vectors. Nature, 348, 91-95.

Mould, J., et al., (2008). A Point-Source Survey of M31 with the Spitzer Space Telescope. ApJ 687 230-241.

Muench, A. et al., (2008). Star Formation in the Orion Nebula I: Stellar Content. In Bo Reipurth, ed. Handbook of Star Forming Regions Vol. I Astronomical Society of the Pacific, 2008

Muno, M. P., et al., (2005). A Lack of Radio Emission from Neutron Star Low-Mass X-Ray Binaries. ApJ 626 1020-1027.

Mushegian, A, (2008). Gene content of LUCA, the last universal common ancestor. Front Biosci. 13:4657-66.

Naeem, S., et al., (2000). Producer-decomposer co-dependency influences biodiversity effects Nature, 403, 762-764.

Naganuma, T., Sekine, Y (2010). Hydrocarbon Lakes and Watery Matrices/Habitats for Life on Titan. Journal of Cosmology, 5. 905-911.

Nagy, B., Meinschein, W. G. Hennessy, D, J. 1961, Mass-spectroscopic analysis of the Orgueil meteorite: evidence for biogenic hydrocarbons. Annals of the New York Academy of Sciences 93, 25-35.

Nagy, B., Claus, G., Hennessy, D, J., 1962, Organic Particles embedded in Minerals in the Orgueil and Ivuna Carbonaceous Chondrites. Nature 193, 1129 - 1133.

Nagy, B., Fredriksson, K., Kudynowkski, J., Carlson, L. 1963a, Ultra-violet Spectra of Organized Elements. Nature 200, 565 - 566.

Nagy, B., Fredriksson, K., Urey, C., Claus, G., Anderson, C. A., Percy, J. 1963b. Electron Probe Microanalysis of Organized Elements in the Orgueil Meteorite, Nature 198, 121 - 125.

Nagy, B., Bitz, M. C. 1963c. Long-chain fatty acids from Orgueil meteorite. Archives of

Biochemistry and Biophysics, 101, 240-263.

Nakabachi A, Yamashita A, Toh H, Ishikawa H, Dunbar HE, Moran NA, Hattori M. (2006). The 160-kilobase genome of the bacterial endosymbiont Carsonella. Science 314:267.

Napier, W. M. (2004). A mechanism for interstellar panspermia. Mon. Not. R. Soc. 348, 46-51.

Nasim, A, James, A. P., (1978). Microbial Life in Extreme Environments. Academic Press.

Natarajan, P., Sigurdsson, S., Silk, J. (1998). MNRAS 298, 577.

Nei, M., Xu, P. & Glazko, G. (2001). Estimation of divergence times from multiprotein sequences for a few mammalian species and several distantly related organisms Proc. Natl. Acad. Sci. USA 98, 2497-2502.

Nelson KE, Clayton RA, Gill SR, Gwinn ML, Dodson RJ, Haft DH, Hickey EK, Peterson JD, Nelson WC, Ketchum KA, et al. (1999). Evidence for lateral gene transfer between Archaea and bacteria from genome sequence of Thermotoga maritima. Nature, 399:323–329.

Nemchin, A. A., Whitehouse, M.J., Menneken, M., Geisler, T., Pidgeon, R.T., Wilde, S. A. (2008). A light carbon reservoir recorded in zircon-hosted diamond from the Jack Hills. Nature 454, 92-95.

Newcomb, W. W., et al, (2001;). The UL6 Gene Product Forms the Portal for Entry of DNA into the Herpes Simplex Virus Capsid J. Virol. 75 , 10923-10932.

Ng, M., & Yanofsky, M. (2001). Function and evolution of the plant MADS-box gene family. Nature Reviews Genetics 2, 186-195.

Nicholson, W. L., Munakata, N., Horneck, G., Melosh, H. J., Setlow, P. (2000). Resistance of Bacillus Endospores to Extreme Terrestrial and Extraterrestrial Environments, Microbiology and Molecular Biology Reviews 64, 548-572.

Nikoh N, Tanaka K, Shibata F, Kondo N, Hizume M, Shimada M, Fukatsu T. (2008) Wolbachia genome integrated in an insect chromosome: evolution and fate of laterally transferred endosymbiont genes. Genome Res. 18:272–280.

Nisbet, E.G, & Nisbet, R.E. (2008). Methane, oxygen, photosynthesis, rubisco and the regulation of the air through time Philos Trans R Soc Lond B Biol Sci. 363, 2745-2754.

Nitschke, W., Russell, M.J. (2010). Just like the universe the emergence of life had high enthalpy and low entropy beginnings. Journal of Cosmology, 10, 3200-3216.

Nixon, J.E.,Wang, A.,Morrison, H.G., McArthur, A.G., Sogin, M.L., Loftus, B.J., Samuelson, J. (2002) A spliceosomal intron in Giardia lamblia. Proc. Natl. Acad. Sci. 99:3701–3705.

Norkin, L. C. (2009). Virology: Molecular Biology and Pathogenesis. ASM Press.

Nosenko, T., & Bhattacharya, D. (2007). Horizontal gene transfer in chromalveolates. BMC Evol. Biol. 7, 173.

Nyquist L. E., Bansal B. M., Wiesmann H., and Shih C.-Y. (1995) Martians young and old: Zagami and ALH84001 (abstract). Lunar Planet. Sci. XXVI, 1065-1066.

O'Brien TW (2002) Evolution of a protein-rich mitochondrial ribosome: Implications for human genetic disease. Gene 286: 73–79.

O'Dell, C. R., Muench, A., Smith, N., Zapata, L. (2008). Star Formation in the Orion Nebula II: Gas, Dust, Proplyds and Outflows. In Bo Reipurth, Ed. Handbook of Star Forming Regions, Volume I: The Northern Sky ASP Monograph Publications.

Olson JM (2006). Photosynthesis in the Archean era. Photosyn. Res. 88 (2): 109–117.

O'Neil, J., Carlson, R. W., Francis, E., Stevenson, R. K. (2008). Neodymium-142 Evidence for Hadean Mafic Crust Science 321, 1828 - 1831.

Pace NR. (2006) Time for a change. Nature 441:289.

Ooosterloo, T.A., Morganti, R. (2005). A&A 429, 469.

O'Neil, J., Carlson, R. W., Francis, E., Stevenson, R. K. (2008). Neodymium-142 Evidence for Hadean Mafic Crust Science 321, 1828 - 1831.

Oparin, A. I. (2003) Origin of Life. Dover.

Osterbrock, D. E., and Ferland, G. J. (2005). Astrophysics Of Gaseous Nebulae And Active

Galactic Nuclei University Science Books.

Pace, G., and Pasquini, L. (2004) The age-activity-rotation relationship in solar-type stars A&A 426 3 (2004) 1021-1034.

Pagaling, E., et al., (2007). Sequence analysis of an Archaeal virus isolated from a hypersaline lake in Inner Mongolia, China. BMC Genomics, 8:410doi:10.1186/1471-2164-8-410.

Panter, B., Jimenez, R., Heavens, A.F., Charlot, S., (2007). MNRAS 378, 1550.

Parkhill, J., et al., (2001). Genome sequence of Yersinia pestis, the causative agent of plague. Nature 413, 523-527.

Parseval, N, de, Heidmann, T. (2005). Human endogenous retroviruses: from infectious elements to human genes Cytogenet Genome Res, 110:318-332.

Pasquin, L., et al., (2005) Early star formation in the Galaxy from beryllium and oxygen abundances Astronomy & Astrophysics 436 3, L57-L60.

Patzke, S., M. Lindeskog, E. Munthe, and H. C. Aasheim. (2002). Characterization of a novel human endogenous retrovirus, HERV-H/F, expressed in human leukemia cell lines.Virology 303:164-173.

Pavlov, A.A., Kasting, J.F, Brown, L.L,, Rages, K.A., Freedman, R. (2000) Greenhouse warming by CH_4 in the atmosphere of early Earth. J. Geophys. Res. 105, 11981–11990.

Pavlov, A.A, Kasting, J.F., Brown, L.L. (2001). UV-shielding of NH_3 and O_2 by organic hazes in the Archean atmosphere. J. Geophys. Res. 106, 23267–23287.

Pavlov, A.A, Hurtgen, M.T, Kasting, J.F, & Arthur, M.A (2003). Methane-rich Proterozoic atmosphere? Geology. 31, 87–90.

Pelaz, S., Ditta, G.S., Baumann, E., Wisman, E., Yanofsky, M.F. (2000). B and C floral organ identity functions require SEPALLATA MADS-box genes. Nature, 405, 200-203.

Pelaz S, et al., (2001). Conversion of leaves into petals in Arabidopsis. Curr Biol. 11, :182-184.

Pereto J, Lopez-Garcia P, Moreira D. (2004). Ancestral lipid biosynthesis and early membrane evolution. Trends Biochem Sci 29:469–477

Perichon B., Bogaerts P., Lambert T., Frangeul L., Courvalin P., Galimand M. (2008). Sequence of conjugative plasmid pIP1206 mediating resistance to aminoglycosides by 16S rRNA methylation and to hydrophilic fluoroquinolones by efflux. Antimicrob. Agents Chemother. 52, 2581–2592.

Perriaud, L., Lachuer, J., Dante, R. (2012) The" Methyl-CpG Binding Domain Protein 2" plays a repressive role in relation to the promoter CpG content in the Normal Human Cell Line MRC5. Current pharmaceutical design

Perry, R. D., and Fetherston, J. D. (1997). Yersinia pestis--etiologic agent of plague. Clin Microbiol Rev. 10, 35-66.

Peterson, KJ., & Butterfield, N.J. (2005). Origin of the eumetazoa: testing ecological predictions of molecular clocks against the proterozoic fossil record. Proc Natl Acad Sci USA, 102, 9547–9552.

Peterson, K. J., et al., (2004). Estimating metazoan divergence times with a molecular clock -PNAS, 101, 6536-6541.

Pflug, H. D. (1978). Yeast-like microfossils detected in oldest sediments of the earth. Journal Naturwissenschaften 65, 121-134.

Pflug, H.D., (1984). Utrafine structure of the organic matter in meteorites, in: N.C. Wickramasinghe, (ed.) Fundamental Studies and the Future of Science, Cardiff: Univ. College Cardiff Press, pp 24-37.

Pflug, H.D. (1984). Microvesicles in meteorites, a model of pre-biotic evolution. Journal Naturwissenschaften, 71, 531-533.

Pflug, H.D. and Heinz, B., 1997. Analysis of fossil-organic nanostructures – terrestrial and extraterrestrial, Proc SPIE, 3111, 86-97.

Pieters, C. M., et al., (2009). Character and Spatial Distribution of OH/H23 on Chandrayaan-1. Science 326 (5952), 568.

Pigliucci, M.(2008) Opinion - Is evolvability evolvable? Nat. Rev. Genet. 9, 75–82

Poccia, N., et al., (2010). The Emergence of Life in the Universe at the Epoch of Dark Energy Domination. Journal of Cosmology, 5. 875-882.

Poduri, A., Evrony, G. D., Cai, Z., & Walsh, C. A. (2013). Somatic Mutation, Genomic Variation, and Neurological Disease, Science, 341,

Poinar, G., & Poinar, R. (2008). What Bugged the Dinosaurs? Insects, Disease and death in the Cretaceus. Princeton University Press. Princeton University Press.

Poitrasson, F., Alexander, N. Hallidaya, N., Leea, D-C. (2004). Sylvain Levasseura and Nadya Teutscha, d Iron isotope differences between Earth, Moon, Mars and Vesta as possible records of contrasted accretion mechanisms. Earth and Planetary Science Letters 223, 253-266.

Polaczyk, P. J., Gasperini, R., & Gibson, G. (1998). Naturally occurring genetic variation affects Drosophilia photoreceptor determination. Developl. Genes Evol. 207, 462-470.

Poleshko, A., Shalginskikh, N. & Katz, R. A. (2012) 3 Functional networks of human epigenetic factors, Epigenomics, From Chromatin Biology to Therapuetics, Appasania, K (Ed). Cambridge U. Press.

Polz, M. F., Aim, E. J., Hanage, W. P. (2013) Horizontal gene transfer and the evolution of bacterial and archaeal population structure, Trends in Genetics, 29, 170-175

Ponferrada VG, Mauck BS, Wooley DP. (2003). The envelope glycoprotein of human endogenous retrovirus HERV-W induces cellular resistance to spleen necrosis virus. Arch Virol; 148:659-675.

Poole A, Penny D. (2007). Eukaryote evolution: engulfed by speculation. Nature, 447:913.

Porter, K., et al., (2007). Virus–host interactions in salt lakes Current Opinion in Microbiology, 10, 18-424.

Porter, S. M. and Knoll, A. H. (2000). Testate amoebae in the Neoproterozoic Era: Evidence from vase-shaped microfossils in the Chuar Group, Grand Canyon. Paleobiology 26, 360-385.

Prangishvili, D., et al., (2006). Unique viral genomes in the third domain of life, Virus Research, 117, 52-67.

Prangishvili D, Forterre P, Garrett RA. (2006). Viruses of the Archaea: a unifying view. Nat Rev Microbiol. 4(11):837-48.

Prasad, S. S., & Tarafdar, S. P. 1983. UV radiation field inside dense clouds. The Astrophysical Journal, 267, 603-609.

Pratt, WB and Toft, D.O (2003). Regulation of signaling protein function and trafficking by the hsp90/hsp70-based chaperone machinery. Exp Biol Med 228:111–133.

Price, P. B., (2000). A habitat for psychrophiles in deep Antarctic ice, Proc. Natl. Acad. Sci. USA, 97:1247-1251.

Prothero, D. R. (1998). Bringing Fossils to Life: An Introduction to Paleobiology. WCB/McGrow-Hill, USA.

Prudhomme, S. Bonnaud, B. Mallet, F. (2005). Endogenous retroviruses and animal reproduction. Retrotransposable Elements and Gene Evolution, 110, 1-4.

Rabosky, D. L. & Adams, D. C. (2012) Rates of morphological evolution are correlated with species richness in salamanders. Evolution 66, 1807–1818

Rabosky, D. L. et al. (2013). Rates of speciation and morphological evolution are correlated across the largest vertebrate radiation. Nature Communications, 1-8.

Rachowicz, L. J., et al. (2006). Emerging infectious diseases as a proximate cause of amphibian mass mortality. Ecology, 87, 1671-1683.

Rafikov, R. (2006). ApJ, 648, 666.

Rampelotto, P. H. (2009). Are We Descendants of Extraterrestrials? Journal of Cosmology, 1, 86-88.

Rampelotto, P. H. (2010). The Search for Life on Other Planets: Sulfur-Based, Silicon-Based, Ammonia-Based Life. Journal of Cosmology, 5. 818-827.

Randles, W. G. L. (1999). The Unmaking of the Medieval Christian Cosmos. Ashgate Publishing.

Rankenburg, K., Brandon, A. D., Neal, C. R. (2006). Neodymium Isotope Evidence for a Chondritic Composition of the Moon Science 312. no. 5778, 1369 - 1372.

Rappaport, L., Oliviero, P., Samuel, J.L. (1998). Cytoskeleton and mitochondrial morphology and function. Mol and Cell Biochem. 184, 101–105.

Raup, D. M. and Sepkoski J. J. (1982). Mass extinctions in the marine fossil record. Science, 215: 1501-1503.

Raup, D. M. (1992) Bad genes or bad luck. Norton, New York.

Rejkuba, M., Minniti, D., Courbin, F., Silva, D.R. (2002). ApJ 564, 688.

Reus, K., et al. (2001). HERV-K(OLD): ancestor sequences of the human endogenous retrovirus family HERV-K(HML-2). J. Virol. 75:8917-8926.

Ribeiro S, Golding G B (1998) The mosaic nature of the eukaryotic nucleus Mol Biol Evol 15:779–788.

Rice. G., et al., (2001). Viruses from extreme thermal environments. PNAS, 98, 13341-13345.

Rice, G., et al., (2004) The structure of a thermophilic archaeal virus shows a double-stranded DNA viral capsid type that spans all domains of life. PNAS 101, 7716-7720.

Richardson, D. J., 2000. Bacterial respiration: a flexible process for a changing environment Microbiology, 146:551-571.

Rigden, J. S. (2003) Hydrogen: The Essential Element. Harvard University Press.

Rivikina, E., et al. (1998). Geomicrobiology,15, 187.

Rivera MC, Lake JA. (1992). Evidence that eukaryotes and eocyte prokaryotes are immediate relatives. Science, 257:74–76.

Rivera, M.C., & Lake, J.A. (2004). The ring of life provides evidence for a genome fusion origin of eukaryotes. Nature, 431, 152–155.

Rivera M C, Rain R, Moore J E, Lake J A (1998) Proc Natl Acad Sci USA 95:6239–6244.

Robertson, E. C. (2001). The Interior of the Earth". USGS. http://pubs.usgs.gov/gip/interior/.

Robertson C, Harris J, Spear J, Pace N (2005). Phylogenetic diversity and ecology of environmental Archaea. Curr Opin Microbiol 8 (6): 638–42.

Rode, O.D. et al., (1979). Atlas of Photomicrographs of the Surface Structures of Lunar Regolith Particles, Boston: D. Reidel Publishing Co.

Rogers MB, Watkins RF, Harper JT, Durnford DG, Gray MW, Keeling PJ., (2007). A complex and punctate distribution of three eukaryotic genes derived by lateral gene transfer. BMC Evol Biol.;7:89.

Rogozin IB, Wolf YI, Sorokin AV, Mirkin BG, Koonin EV. (2003). Remarkable inter-kingdom conservation of intron positions and massive, lineage-specific intron loss and gain in eukaryotic evolution. Curr Biol. Sep 2;13(17):1512-7.

Romancer, M. Le, et al., (2007). Viruses in extreme environments, Reviews in Environmental Science and Biotechnology, 6, 17-31.

Romano,, C. M. et al., (2007). Demographic Histories of ERV-K in Humans, Chimpanzees and Rhesus Monkeys PLoS ONE. 2(10): e1026.

Romano, C. M., et al., (2008). -Journal of Molecular Evolution, 66, 292-297.

Roscoe, S.M. (1969). Huronian rocks and uraniferous conglomerates in the Canadian Shield. Geol. Surv. Can. Pap. 8, 68-40.

Roscoe, S.M. (1973). The Huronian Supergroup: a Paleophebian succession showing evidence of atmospheric evolution. Geol. Soc. Can. Spec. Pap. 12, 31–48.

Rosing, M. T. (1999). C-13-depleted carbon microparticles in > 3700-Ma sea-floor sedimentary rocks from west Greenland. Science 283, 674-676.

Rosing, M. T., Frei, R. (2004). U-rich Archaean sea-floor sediments from Greenland -

indications of > 3700 Ma oxygenic photosynthesis. Earth and Planetary Science Letters 217, 237-244.

Roy, S. W. (2003). Recent evidence for the exon theory of genes. Genetica 118, 251–266.

Roy, S. W., (2004). The origin of recent introns: transposons? Genome Biology, 5:251.

Roy, S. W. (2006). Intron-rich ancestors. Trends Genet. 2006 Sep;22(9):468-71. Epub

Roy SW, Gilbert W. (2006). The evolution of spliceosomal introns: patterns, puzzles and progress. Nat. Rev. Genet. 7:211–221.

Roy, S. W., Fedorov, A. & Gilbert, W. (2003) Large-scale comparison of intron positions in mammalian genes shows intron loss but no gain.Proc. Natl. Acad. Sci. USA 100:, 7158–7162.

Roy, S. W., Lewis, B. P., Fedorov, A. & Gilbert, W. (2001). Footprints of primordial introns on the eukaryotic genome. Trends Genet. 17, 496–498.

Roy, S. W., Nosaka, M., de Souza, S. J. & Gilbert, W. (1999). Centripetal modules and ancient introns. Gene 238, 85–91.

Rozanov, A. Yu and Hoover, R.B., 2003. Atlas of bacteriomorphs in carbonaceous chondrites, ProcSPIE, 5163, 23-35.

Ruddimann, W.F., (2005). Plows, Plagues, and Petroleum: How Humans Took Control of Climate, Princeton University Press.

Russell, M. J., and Arndt, N. T. (2005). Geodynamic and metabolic cycles in the Hadean. Biogeosciences, 2, 97-111.

Russell, M.J., and Hall, A. J. (1999). On the inevitable emergence of life on Mars. In: Hiscox, J.A. (Ed.), The Search for Life on Mars, Proceedings of the 1st UK Conference, British Interplanetary Society, London, pp. 26-36.

Russell, M.J., Kanik, I. (2010). Why does life start, What does It do, where might it be, how might we find it? Journal of Cosmology, 5, 1008-1039.

Rutherford, S.L. (2003). Between genotype and phenotype: protein chaperones and evolvability. Nat Rev Genet 4:263–274.

Rutherford, S. L., & Lindquist, S. (1998). Hsp90 as a capacitor for morphological evolution. Nature 396, 336-342.

Sackmann, I. J.; Boothroyd, A. I.; Kraemer, K. E. (1993). Our Sun. Past, Present and Future". Astrophysical Journal 418: 417-488

Sahl. J. W., et al., 2008. Subsurface Microbial Diversity in Deep-Granitic-Fracture Water in Colorado Applied and Environmental Microbiology, 74, 143-152.

Saltzman, M. R., et al. (1995). Sea-level-driven changes in ocean chemistry at an Upper Cambrian extinction horizon. Geology, 23, 893-896.

Sancho L. G., de la Torre, R., Horneck, G., Ascaso, C. , de los Rios, A. Pintado,A., Wierzchos, J.,Schuster, M. 2007. Lichens Survive in Space: Results from the 2005 LICHENS Experiment Astrobiology. 7, 443-454.

Sandford S. A., et al. (2006). Organics captured from Comet 81P/ Wild 2 by the Stardust spacecraft. Science 314(5806):1720– 1724.

Sangster, T.A, et al., (2004). Under cover: causes, effects and implications of Hsp90-mediated genetic capacitance. BioEssays 26:348–62.

Sato, B., Fischer, D. A., Henry, G.W. et al. (2005), ApJ, 633, 465.

Saumon, D., Hubbard, W. B., Burrows, A., Guillot, T., Lunine, J. I., Chabrier, G. (1996), ApJ, 460, 993.

Schafer, G., Purschke, W. & Schmidt, C. L. (1996). On the origin of respiration: electron transport proteins from archaea to man.FEMS Microbiol Rev 18, 173-188.

Scheifele, L. Z., Boeke, J. D. (2008). From the shards of a shattered genome, diversity. Proc Natl Acad Sci 105, 11593–11594.

Schoenberg, R., Kamber, B.S., Collerson, K.D., Moorbath, S. 2002. Tungsten isotope evidence from approximately 3.8-Gyr metamorphosed sediments for early meteorite bombardment of the Earth. Nature 418, 403-405.

Alien Viruses, Genetics, Cambrian Explosion: Humans

Schneiker S, Perlova O, Kaiser O, Gerth K, Alici A, Altmeyer MO, Bartels D, Bekel T, Beyer S, Bode E, et al. (2007) Complete genome sequence of the myxobacterium Sorangium cellulosum. Nat. Biotechnol. 25:1281–1289.

Schulze-Makuch, D.(2010). Io: Is Life Possible Between Fire and Ice? Journal of Cosmology, 5, 833-842.

Schulze-Makuch, D., Grinspoon, D. H., Abbas, O., Irwin, L. N., Bullock, M.A. (2004). A sulfur-based survival strategy for putative phototrophic life in the Venusian atmosphere. Astrobiology, 4, 11-18.

Schulze-Makuch, D., Irwin, L. N. (2006). The prospect of alien life in exotic forms on other worlds. Naturwissenschaften, 93, 155-172.

Schwartzman D, Caldeira K, Pavlov A. (2008). Cyanobacterial emergence at 2.8 gya and greenhouse feedbacks.: Astrobiology, 8, 187-203.

Schwegler E., et al., (2001). Ph. Rev. Letter, 87, 265501.

Sears DW, Kral TA (1998).Martian "microfossils" in lunar meteorites? Meteorit Planet Sci. 33, 791-4

Seifarth, W., et al., (1998). Proviral structure, chromosomal location, and expression of HERV-K-T47D, a novel human endogenous retrovirus derived from T47D particles. J. Virol. 72:8384-8391.

Seifarth, W., et al., (2005). Comprehensive Analysis of Human Endogenous Retrovirus Transcriptional Activity in Human Tissues with a Retrovirus-Specific Microarray Journal of Virology, 79, 341-352.

Seleme MC, Vetter MR, Cordaux R, Bastone L, Batzer MA, Kazazian HH Jr. (2006) Extensive individual variation in L1 retrotransposition capability contributes to human genetic diversity. Proc Natl Acad Sci USA 103: 6611Y6616.

Setlow B; Setlow P. (1995). Binding to DNA protects alpha/beta-type, small, Journal of bacteriology 177(14):4149-51.

Setlow, B., Setlow, P. (1995). Small, acid-soluble proteins bound to DNA protect Bacillus subtilis spores from killing by dry heat. Appl Environ Microbiol. 61, 2787–2790.

Shabalina, S. A. et al., (2010). Distinct patterns of expression and evolution of intronless and intron-containing mammalian genes, Mol Biol Evol, 27, 1745-1749

Sharov, A.A. (2009).Exponential Increase of Genetic Complexity Supports Extra-Terrestrial Origin of Life. Journal of Cosmology, 1, 63-65.

Sharov, A. A. (2010). Genetic Gradualism and the ExtraTerrestrial Origin of Life. Journal of Cosmology, 5,

Sharov, A. A. & Gordon, R. (2013) Life Before Earth, arxiv 1304.3381

Sheehan, P. W., (2001). The late ordovician mass extinction. Annual Review of Earth and Planetary Sciences, 29, 331-364

Shen, B. (2008). Global Anoxia and mass extinction at the Cenomanian-Turonian boundary triggered by subduction zone volcanism. 2008 Joint Annual Meeting, Celebrating the International Year of Planet Earth, October 5-9, Houston, Texas.

Sharp, P. A. (1991). Five easy pieces. Science 254, 663

Sherman LA, Pauw P., (1976). Infection of Synechococcus cedrorum by the cyanophage AS-1M. II. Protein and DNA synthesis.Virology. 71(1):17-27.

Shi, Y-B (1999) Amphibian Metamorphosis: From Morphology to Molecular Biology, Wiley-Liss.

Sidharth, B. G. (2009). In defense of abiogenesis, Journal of Cosmology, 1, 73-75.

Silk, J. (2005). MNRAS 364, 1337.

Silk, J., Norman, C. (2009). ApJ 700, 262 Feain, I.J., Papadopoulos, P.P., Ekers, R.D., Middelberg, E. 2007, ApJ 662, 872.

Simpson, A.G., MacQuarrie, E.K., Roger, A.J. (2002) Eukaryotic evolution: Early origin of canonical introns. Nature 419:270

Slater FR, Bailey MJ, Tett AJ, Turner SL. (2008). Progress towards understanding the fate of plasmids in bacterial communities. FEMS Microbiol. Ecol.

Sleep, N. H., Bird, D. K. (2008). Evolutionary ecology during the rise of dioxygen in the Earth's atmosphere--Phil. Trans. R. Soc. B 27, vol. 363 no. 1504 2651-2664.

Sletvold H., Johnsen P.J., Hamre I., Simonsen G.S., Sundsfjord A., Nielsen K.M. (2008). Complete sequence of Enterococcus faecium pVEF3 and the detection of an omega epsilon zeta toxin-antitoxin module and an ABC transporter. Plasmid. 60, 75–85.

Smith, K. F. et al. (2006) Evidence for the Role of Infectious Disease in Species Extinction and Endangerment. Conservation Biology, 20, 1349–1357.

Smith M., J, Smith N H, O'Rourke M, Spratt B G (1993). How clonal are bacteria? Proc Natl Acad Sci USA 90:4384–4388.

Snel B, Bork P, Huynen MA. (2002) Genomes in flux: the evolution of archaeal and proteobacterial gene content. Genome Res. 12:17–25.

Sodergren, E,, et. al (2007). The genome of the sea urchin Strongylocentrotus purpuratus. Science, 314, 941-52.

Soffen, G.A. 1965. NASA Technical Report, N65-23980.

Srivastava M, Begovic E, Chapman J, Putnam NH, Hellsten U, Kawashima T, Kuo A, Mitros T, Salamov A, Carpenter ML, Signorovitch AY, Moreno MA, Kamm K, Grimwood J, Schmutz J, Shapiro H, Grigoriev IV, Buss LW, Schierwater B, Dellaporta SL, Rokhsar DS. (2008). The Trichoplax genome and the nature of placozoans, Nature 454, 955-960.

Stephens, D. and Noll, K. (2006). Astronomical Journal, 131, 1142. Cruz, K. et al, 2009, AJ, 137, 3345

Strachan, T., & Read, A. (1996). Human Molecular GeneticsBios Scientific Publishers Ltd.

Stankiewicz P, Lupski JR. (2010). Structural variation in the human genome and its role in disease. Annual Review of Medicine, 61, 437-455.

Steffen, W, Crutzen, P.J., McNeill, J.R. (2007). The Anthropocene: Are Humans Now Overwhelming the Great Forces of Nature? Ambio, 36, 614-621.

Stokes H.W., . Hall R.M. (1989). A novel family of potentially mobile DNA elements encoding site-specific gene-integration functions: integrons. Mol. Microbiol. 3, 1669–1683.

Stoltzfus, A. (1999). On the possibility of constructive neutral evolution. J. Mol. Evol. 49, 169–181.

Syvanen, M., (1985). Cross-species Gene Transfer; Implications for a New Theory of Evolution, J.theor. Biol. 112, 333—343.

Syvanen M, Kado CI, eds. Horizontal Gene Transfer (2002) San Diego: Academic Press.

Sullivan MB, Coleman ML, Weigele P, Rohwer F, Chisholm SW. (2005). Three Prochlorococcus cyanophage genomes: signature features and ecological interpretations; PLoS Biol. 3(5):e144.

Sullivan MB, Coleman ML, Weigele P, Rohwer F, Chisholm SW. (2005). Three Prochlorococcus cyanophage genomes: signature features and ecological interpretations; PLoS Biol. 2005 May;3(5):e144.

Sullivan M.B, et al., (2006). Prevalence and evolution of core photosystem II genes in marine cyanobacterial viruses and their hosts. PLoS Biol. 4(8):e234.

Sullivan, N.J, Geisbert TW, Geisbert JB, Shedlock DJ, Xu L, et al. (2006) Immune Protection of Nonhuman Primates against Ebola Virus with Single Low-Dose Adenovirus Vectors Encoding Modified GPs. PLoS Med 3(6): e177. doi:10.1371/journal.pmed.0030177.

Sullivan, R., Fassolitis, A.C., Larkin, E.P., Read, R.B. and Peeler, J.T. (1971) Inactivation of thirty viruses by gamma radiation, Appl. Microbiology, 22, 61-65.

Summons R.E, Bradley A.S, Jahnke L.L, Waldbauer R (2006). Steroids, triterpenoids and molecular oxygen. Phil. Trans. R. Soc. B. 361, 951–968.

Summons, R.E. et al. (1999) Nature, 400 : 554-557.

Sun, S.-S. (1982). Chemical composition and origin of the earth's primitive mantle. Geochimica et Cosmochimica Acta, vol. 46, Feb. 1982, p. 179-192.

Sun, S-S, and Nesbitt, R. W. (1977) Chemical heterogeneity of the Archaean mantle, composition of the earth and mantle evolution Earth and Planetary Science Letters, Volume 35, Issue 3, p. 429-448.

Sunshine, J. M. et al., (2009). Temporal and Spatial Variability of Lunar Hydration As Observed by the Deep Impact Spacecraft. Science 326 (5952), 565.

Sverdlov ED. (2000). Retroviruses and primate evolution. Bioessays 22:161–171.

Sundin, GW.(2007). Genomic insights into the contribution of phytopathogenic bacterial plasmids to the evolutionary history of their hosts. Annu. Rev. Phytopathol. 45:129–151.

Sung, W. K., et al. (2006). Neoproterozoic Bimodal Volcanism in the Okcheon Belt, South Korea, and Its Comparison with the Nanhua Rift, South China: Implications for Rifting in Rodinia, The Journal of Geology, 114, 717–733.

Sverdlov ED. (2000). Retroviruses and primate evolution. Bioessays 22:161–171.

Swift, M.J., et al., 1979). Decomposition in Terrestrial Ecosystems. University of California Press, Berkeley, CA.

Tamae C., et al., (2008). Determination of antibiotic hypersensitivity among 4000 single-gene-knockout mutants of Escherichia coli. J. Bacteriol. 190, 5981–5988.

Tamames, J., et al., (2007). The frontier between cell and organelle: genome analysis of Candidatus Carsonella ruddii BMC Evolutionary Biology, 7, 181.

Tanner, L.H., Lucas, S.G., & Chapman, M.G. (2004). Assessing the record and causes of Late Triassic extinctions. Earth-Science Reviews, 65, 103-139.

Thomas-Keprta, K. L., et al., (2009). Origins of magnetite nanocrystals in Martian meteorite ALH84001. Geochimica et Cosmochimica Acta, 73, 6631-6677.

Thompson, S. L. and Crutzen, P. J. (1988). Acute Effects of a Large Bolide Impact Simulated by a Global Atmospheric Circulation Model. Topical Conference on Global Catastrophes in Earth History: An Interdisciplinary Conference on Impacts, Volcanism, and Mass Mortality. October 20-23, 1988, Snowbird, Utah.

Throop, H.; Bally, J. (2009). UV Photolysis and Creation of Complex Organic Molecules in the Solar Nebula. 40th Lunar and Planetary Science Conference, (Lunar and Planetary Science XL), held March 23-27, 2009 in The Woodlands, Texas, id.2139.

Timmis, JN, et al. (2004). Endosymbiotic gene transfer: organelle genomes forge eukaryotic chromosomes Nature Reviews Genetics 5, 123-135

Tomarev, S. I., (1997). Pax-6, eyes absent, and Prox 1 in eye development. Int. J. Dev. Biol. 41: 835 - 842.

Tomarev, S. I., et al., (1996). Chicken homeobox gene Prox 1 related to Drosophila prosperois expressed in the developing lens and retina. Dev. Dynamics 206: 354-377. Thomas C.M., Nielsen K.M. (2005). Mechanisms of, and barriers to, horizontal gene transfer between bacteria. Nat. Rev. Microbiol. 3, 711–721.

Torosian, S. D., et al., (2009). A refrigeration temperature of 4 degrees C does not prevent static growth of Yersinia pestis in heart infusion broth. Can J Microbiol. 2009 Sep ;55 (9):1119-24 19898555

Tovar J, Fischer A, Clark CG.(1999). The mitosome, a novel organelle related to mitochondria in the amitochondrial parasite Entamoeba histolytica. Molecular Microbiology 32, 1013–1021.

Theissen, G. et al., (2000). A short history of MADS-box genes in plants Plant Molecular Biology, 42, 115-149.

Trilling, D. and Bernstein, G. (2006). Astronomical Journal, 131, 1149.

Tully, R.B. (1986). Astrophys. J. 303, 25-38.

Turcotte, D. L., Schubert, G. (2002). Geodynamics. Cambridge, England, UK: Cambridge University Press.

Twitchett, R. J. (2006). The palaeoclimatology, palaeoecology and palaeoenvironmental analysis of mass extinction events. Palaeogeogr., Palaeoclimat., Palaeoecol., 232 (2006): 190-213.

Valentine, J. W., Jablonski, D. and Erwin, D. H. (1999). Fossils, molecules and embryos: New perspectives on the Cambrian explosion. Development, 126, 851-859.

Valley,J. W., et al. (2002). A Cool Early Earth,Geology. 30, 351-354.

van de Lagemaat L. N. Josette-Renée Landry1, Dixie L. Mager1, and Patrik Medstrand, (2003). Transposable elements in mammals promote regulatory variation and diversification of genes with specialized functions Trends in Genetics, 19, 530-536.

Vanacova, S., Yan, W., Carlton, J.M., Johnson, P.J. (2005) Spliceosomal introns in the deep-branching eukaryote Trichomonas vaginalis. Proc. Natl. Acad. Sci. 102:4430–4435.

van der Giezen M, Tovar J. (2005) Degenerate mitochondria. EMBO Rep 6:525–530.

Vargas, M., Kashefi, K., Blunt-Harris, E. L. & Lovley, D. R. (1998). Microbiological evidence for Fe(III) reduction on early Earth.Nature 395, 65-67.

Vishwanath P, Favaretto P, Hartman H, Mohr SC, Smith TF. (2004). Ribosomal protein-sequence block structure suggests complex prokaryotic evolution with implications for the origin of eukaryotes. Mol Phylogenet Evol. 33:615–625.

Vogt, P. K. (1997). Historical introduction to the general properties of retroviruses, p. 1-25. In J. M. Coffin, S. H. Hughes, and H. E. Varmus (ed.), Retroviruses. Cold Spring Harbor Laboratory Press, New York, N.Y.

Volff JN, Korting C, Meyer A, Schartl M (2001) Evolution and discontinuous distribution of Rex3 retrotransposons in fish. Mol Biol Evol 18: 427Y431.

Volff JN, Bouneau L, Ozouf-Costaz C, Fischer C (2003) Diversity of retrotransposable elements in compact pufferfish genomes. Trends Genet 19: 674Y678.

Vreeland, R.H., Rosenzweig, W.D., Powers, D.W. (2000). Isolation of a 250 million-year-old halotolerant bacterium from a primary salt crystal. Nature, 407, 897-900.

Wachter, A., Winters, J. M., Schroder, K.-P., Sedlmayr, E. (2008). Dust-driven winds and mass loss of C-rich AGB stars with subsolar metallicities Astronomy & Astrophysics, 1-9.

Wade, M., Johnson, N.A., Jones, R., Siguel, V., & McNaughton, M. (1997). Genetic variation segregating in natural populations of Tribolium castaneum affecting traits observed in hybrids with T. fremani. Genetics, 147, 1235-1247.

Wadhwa, M., Lugmair G. W. (1996). The formation age of carbonates in ALH 84001. Meteoritics, 31, A145.

Wald, R. M. (1992). Space, Time, and Gravity: The Theory of the Big Bang and Black Holes. University of Chicago Press.

Walker, B. J. R. (1970). Viruses respond to environmental exposure (Viruses response to environmental exposure emphasizing temperature, humidity, light and extraterrestrial conditions). JOURNAL OF ENVIRONMENTAL HEALTH. 32, 39-54.

Wang, D Y, Kumar, S., Hedges, S., (1999). Divergence time estimates for the early history of animal phyla and the origin of plants, animals and fungi. Proc Biol Sci. 266, 163–171.

Wang-Johanning, F., et al., (2001). Expression of human endogenous retrovirus K envelope transcripts in human breast cancer.Clin. Cancer Res. 7:1553-1560.

Wang-Johanning, F., et al., (2003). Detecting the expression of human endogenous retrovirus E envelope transcripts in human prostate adenocarcinoma.Cancer 98:187-197.

Wang, J., and Li, Z. H. (2002). History of Neoproterozoic rift basins in South China: implications for Rodinia break-up. Precambrian Res. 122:141–158.

Wang, X.W., Gibson, M.K., Vermeulen, W., Yeh, H., Forrester, K., Sturzbecher, H.W. Hoeijmakers, J.H., & Harris, C.C. (1995). Abrogation of p53-induced Apoptosis by the Hepatitis B Virus X Gene. Cancer Res. 24, 6012–6016.

Ward, P. (2006). Out of Thin Air: Dinosaurs, Birds, and Earth's Ancient Atmosphere. Joseph Henry Press, 296 pp.

Ward, P. (2009). The Medea Hypothesis: Is Life on Earth Ultimately Self- Destructive?

Princeton University Press, Princeton, NJ.

Ward, P. D., et al. (2004). Isotopic evidence bearing on Late Triassic extinction events. Earth and Planetary Science Letters, 224, 589-600.

Waters E, Hohn MJ, Ahel I, Graham DE, Adams MD, Barnstead M, Beeson KY, Bibbs L, Bolanos R, Keller M, et al. (2003). The genome of Nanoarchaeum equitans: insights into early archaeal evolution and derived parasitism. Proc. Natl Acad. Sci. USA 100:12984–12988.

Werner, M. W.; Sellgren, K.; Livingston, J. (2009) The Uniformity of Hydrocarbon Emission from Bright Reflection Nebulae. Bulletin of the American Astronomical Society, Vol. 41, p.219.

Whittington, H. B. (1979). Early arthropods, their appendages and relationships. In M. R. House (Ed.), The origin of major invertebrate groups (pp. 253–268). The Systematics Association Special Volume, 12. London: Academic Press.

Wickham, M. E. et al., (2007). Virulence Is Positively Selected by Transmission Success between Mammalian Hosts. Curr Biol. 2007 Apr 17;: 17442572

Wigler, M., Sweet, R., Sim, G. K., et al. (1979). Transformation of mammalian cells with genes from prokaryotes and eurkarotes. Cell, 16, 777-785.

Wignall, P. B. and Twitchett, R. J. (1996). Oceanic anoxia and the End Permian mass extinction. Science, 272 (5265): 1155 – 1158.

Wilcove, D. S., Rothstein, D., Dubow, D., Phillips, A. and Losos, E. (1998). Quantifying Threats to imperiled species in the United States. BioScience, 48, 2-22.

Wilde, S. A.; Valley, J. W.; Peck, W. H. & Graham, C. M. (2001), "Evidence from detrital zircons for the existence of continental crust and oceans on the Earth 4.4 Gyr ago", Nature 409 (6817): 175–178.

Williams BA, Hirt RP, Lucocq JM, Embley TM. (2002). A mitochondrial remnant in the microsporidian Trachipleistophora hominis. Nature 418:865-869.

Williams, D.A., Brown, W.A., Price, S.D., Rawlings, J.M.C., and Viti, S. (2007). Molecules, ices and astronomy, Astronomy and Geophysics, 48, 25.

Williams R.J.P, Fraústo da Silva J.J.R, (1996). The natural selection of the chemical elements—the environment and life's chemistry. In Clarendon Press, Oxford, UK:Clarendon Press.

Williams R.J.P, Fraústo da Silva J.J.R (2006). The chemistry of evolution: the development of our ecosystem. Elsevier Amsterdam, The Netherlands:Elsevier

Williamson, S. J., et al., (2008). The Sorcerer II Global Ocean Sampling Expedition: Metagenomic Characterization of Viruses within Aquatic Microbial Samples. PLoS ONE. 2008; 3(1): e1456.

Wirstrum. E. S., et al., (2007) A search for pre-biotic molecules in hot cores. A&A 473, 177-180.

Woese, C. (1968). The Genetic Code. Harper & Row.

Woese, C. R. (1987) Bacterial evolution. Microbiol. Rev. 51, 221-271.

Woese, C.R. (2004). A new biology for a new century. Microbiol. Mol. Biol. Rev. 68 (2): 173–86.

Woese, C. R., and Fox, G. E. (1977) Proc. Natl. Acad. Sci. . 74, 5088-5090.

Wolf, Y. I. & Koonin, E. V. (2013). Genome reduction as the dominant mode of evolution, BioEssays, 35, 829-837.

Wollman EL, Jacob F, Hayes W. (1956). Conjugation and genetic recombination in Escherichia coli K-12. Cold Spring Harb. Symp. Quant. Biol.21:141–162.

Wright G.D. (2007). The antibiotic resistome: the nexus of chemical and genetic diversity. Nat. Rev. Microbiol. 5, 175–186.

Xiao, S. H. and Knoll, A. H. (1999). Fossil preservation in the Neoproterozoic Doushantuo phosphorite Lagerstatte, South China. Lethaia 32, 219-240.

Yockey, H.P. (1977). A calculation of the probability of spontaneous biogenesis by information theory. Journal of Theoretical Biology, 67, 377-398.

Yutin, N., et al., (2008). The Deep Archaeal Roots of Eukaryotes Molecular Biology and Evolution 25(8):1619-1630

Zaret, K Palozola, K., Caravaca, JM (2013) Activating the genome during development and exit from mitosis, Epigenetics Chromatin. 2013; 6(Suppl 1): O35.

Zauberman N, Mutsafi Y et al. (2008) PLoS Biology Vol. 6, No. 5, e114 doi:10.1371/journal.pbio.0060114.

Zeidner G, Bielawski JP, Shmoish M, Scanlan DJ, Sabehi G, et al. (2005) Potential photosynthesis gene recombination between Prochlorococcus and Synechococcus via viral intermediates. Environmental Microbiology. 7:1505–1513.

Zelik, M. (2002). Astronomy: The Evolving Universe, Cambridge University Press, Cambridge.

Zhmur, S. I., Gerasimenko, L. M. (1999). Biomorphic forms in carbonaceous meteorite Alliende and possible ecological system - producer of organic matter hondrites" in Instruments, Methods and Missions for Astrobiology II, RB. Hoover, Editor, Proceedings of SPIE Vol. 3755 p. 48-58.

Zhmur, S. I., Rozanov, A. Yu., Gorlenko, V. M. (1997). Lithified remnants of microorganisms in carbonaceous chondrites, Geochemistry International, 35, 58-60.

Zhou, C., Brasier, M. D. and Xue, Y. (2001). Three-dimensional phosphatic preservation of giant acritarchs from the Terminal Proterozoic Doushantuo Formation in Guizhou and Hubei Provinces, South China. Palaeontology 44, 1157-1178.

Zhou, M.-F., et al. (2002). SHRIMP U-Pb zircon geochronological and geochemical evidence for Neoproterozoic arcmagmatism along the western margin of the Yangtze Block, South China. Earth Planet. Sci. Lett. 196, 51– 67.

Zhmur, S. I., Gerasimenko, L. M. (1999). Biomorphic forms in carbonaceous meteorite Alliende and possible ecological system - producer of organic matter hondrites" in Instruments, Methods and Missions for Astrobiology II, RB. Hoover, Editor, Proceedings of SPIE Vol. 3755 p. 48-58.

Zhmur, S. I., Rozanov, A. Yu., Gorlenko, V. M. 1997. Lithified remnants of microorganisms in carbonaceous chondrites, Geochemistry International, 35, 58–60.

Zhuravlev, A. Y., & Wood, R. A. (1996). Anoxia as the cause of the mid-Early Cambrian (Botomian) extinction event. Geology, 24, 311-314.

www.ingramcontent.com/pod-product-compliance
Lightning Source LLC
Chambersburg PA
CBHW081503200326
41518CB00015B/2364